2nd Edition

by Andrie de Vries and
Joris Meys

R For Dummies®, 2nd Edition

Published by: **John Wiley & Sons, Inc.,** 111 River Street, Hoboken, NJ 07030-5774, www.wiley.com

Copyright © 2015 by John Wiley & Sons, Inc., Hoboken, New Jersey

Media and software compilation copyright © 2015 by John Wiley & Sons, Inc. All rights reserved.

Published simultaneously in Canada

For general information on our other products and services, please contact our Customer Care Department within the U.S. at 877-762-2974, outside the U.S. at 317-572-3993, or fax 317-572-4002. For technical support, please visit www.wiley.com/techsupport.

Wiley publishes in a variety of print and electronic formats and by print-on-demand. Some material included with standard print versions of this book may not be included in e-books or in print-on-demand. If this book refers to media such as a CD or DVD that is not included in the version you purchased, you may download this material at http://booksupport.wiley.com. For more information about Wiley products, visit www.wiley.com.

Library of Congress Control Number: 2015941928

ISBN 978-1-119-05580-8 (pbk); ISBN 978-1-119-05583-9 (epub); 978-1-119-05585-3 (epdf)

Manufactured in the United States of America

10 9 8 7 6 5 4 3 2 1

Table of Contents

Introduction

*W*elcome to *R For Dummies,* the book that helps you learn the statistical programming language R quickly and easily.

We can't guarantee that you'll be a guru if you read this book, but you should be able to

- ✔ Perform data analysis by using a variety of powerful tools.
- ✔ Use the power of R to do statistical analysis and data-processing tasks.
- ✔ Appreciate the beauty of using vector-based operations (rather than loops) to do speedy calculations.
- ✔ Appreciate the meaning of the following line of code:

```
knowledge <- apply(theory, 1, sum)
```

- ✔ Know how to find, download, and use code that has been contributed to R by its very active community of developers.
- ✔ Know where to find extra help and resources to take your R coding skills to the next level.
- ✔ Create beautiful graphs and visualizations of your data.

About This Book

R For Dummies is an introduction to the statistical programming language known as R. We start by introducing the interface and work our way from the very basic concepts of the language through more sophisticated data manipulation and analysis.

We illustrate every step with easy-to-follow examples. This book contains numerous code snippets, several write-it-yourself functions you can use later on, and complete analysis scripts. All these are for you to try out yourself.

We don't attempt to give a technical description of how R is programmed internally, but we do focus as much on the why as on the how. R has many features that may seem surprising at first, so we believe it's important to explain both how you should talk to R, and how the R engine interprets what you say. After reading this book, you should be able to manipulate your data in the form you want and understand how to use functions we *didn't* cover in the book (as well as the ones we do cover).

This book is a reference. You don't have to read it from beginning to end. Instead, you can use the table of contents and index to find the information you need. We cross-reference other chapters where you can find more information.

Changes in the Second Edition

Since the publication of the first edition, R has kept evolving and improving. To keep the book accurate, we updated the code to reflect any changes in the latest version of R (version 3.2.0). With the feedback from readers, students, and colleagues we could rework some sections to clarify issues and correct inaccuracies. For example, we modified the code to use double quotes instead of single quotes when using text strings. We also refer to the fundamental units of lists as components, rather than elements.

The new `rfordummies` package contains code examples in the book. Read all about it in Appendix B.

R and RStudio

R For Dummies can be used with any operating system that R runs on. Whether you use Mac, Linux, or Windows, this book will get you on your way with R.

R is more a programming language than an application. When you download R, you automatically download a console application that's suitable for your operating system. However, this application has only basic functionality, and it differs to some extent from one operating system to the next.

RStudio is a cross-platform application, also known as an Integrated Development Environment (IDE) with some very neat features to support R. In this book, we don't assume you use any specific console application. However, RStudio provides a common user interface across the major operating systems. For this reason, we use RStudio to demonstrate some of the concepts rather than any specific operating-system version of R.

Conventions Used in This Book

Code snippets appear like this example, where we simulate 1 million throws of two six-sided dice:

```
> set.seed(42)
> throws <- 1e6
> dice <- replicate(2,
+                      sample(1:6, throws, replace = TRUE)
+ )
> table(rowSums(dice))

     2       3       4       5       6       7       8
 28007   55443   83382  110359  138801  167130  138808
     9      10      11      12
110920   83389   55816   27945
```

Each line of R code in this example is preceded by one of two symbols:

- ✔ **>:** The prompt symbol, >, is not part of your code, and you should not type this when you try the code yourself.

- ✔ **+:** The continuation symbol, +, indicates that this line of code still belongs to the previous line of code. In fact, you don't have to break a line of code into two, but we do this frequently, because it improves the readability of code and helps it fit into the pages of a book.

Lines that start without either the prompt or the continuation symbol are output produced by R. In this case, you get the total number of throws where the dice added up to the numbers 2 through 12. For example, out of 1 million throws of the dice, on 28,007 occasions the numbers on the dice added to 2.

You can copy these code snippets and run them in R, but you have to type them exactly as shown. There are only three exceptions:

- ✔ Don't type the prompt symbol, >.

- ✔ Don't type the continuation symbol, +.

- ✔ Where you put spaces or tabs isn't critical, as long as it isn't in the middle of a keyword. Pay attention to new lines, though.

Instructions to type code into the R console has the > symbol to the left:

```
> print("Hello world!")
```

If you type this into a console and press Enter, R responds with:

```
[1] "Hello world!"
```

For convenience, we collapse these two events into a single block, like this:

```
> print("Hello world!")
[1] "Hello world!"
```

Functions, arguments, and other R keywords appear in `monofont`. For example, to create a plot, you use the `plot()` function. Function names are followed by parentheses — for example, `plot()`. We don't add arguments to the function names mentioned in the text, unless it's really important.

On some occasions we talk about menu commands, such as File⇨Save. This just means that you open the File menu and choose the Save option.

What You're Not to Read

You can use this book however works best for you, but if you're pressed for time (or just not interested in the nitty-gritty details), you can safely skip anything marked with a Technical Stuff icon. You also can skip sidebars (text in gray boxes); they contain interesting information, but nothing critical to your understanding of the subject at hand.

Foolish Assumptions

This book makes the following assumptions about you and your computer:

- **You know your way around a computer.** You know how to download and install software. You know how to find information on the Internet and you have Internet access.

- **You're not necessarily a programmer.** If you are a programmer, and you're used to coding in other languages, you may want to read the notes marked by the Technical Stuff icon — there, we fill you in on how R is similar to, or different from, other common languages.

- **You're not a statistician, but you understand the very basics of statistics.** *R For Dummies* isn't a statistics book, although we do show you how to do some basic statistics using R. If you want to understand the statistical stuff in more depth, we recommend *Statistics For Dummies*, 2nd Edition, by Deborah J. Rumsey, PhD (Wiley).

- **You want to explore new stuff.** You like to solve problems and aren't afraid of trying things out in the R console.

How This Book Is Organized

The book is organized in six parts. Here's what each of the six parts covers.

Part I: Getting Started with R Programming

In this part, you write your first script. You use the powerful concept of vectors to make simultaneous calculations on many variables at once. You work with the R workspace (in other words, how to create, modify, or remove variables). You find out how to save your work and retrieve and modify script files that you wrote in previous sessions. We also introduce some fundamentals of R (for example, how to install packages).

Part II: Getting Down to Work in R

In this part, we fill you in on the three R's: reading, 'riting, and 'rithmetic — in other words, working with text and numbers (and dates for good measure). You also get to use the very important data structures of *lists* and *data frames*.

Part III: Coding in R

R is a programming language, so you need to know how to write and understand functions. In this part, we show you how to do this, as well as how to control the logic flow of your scripts by making choices using if statements, as well as looping through your code to perform repetitive actions. We explain how to make sense of and deal with warnings and errors that you may experience in your code. Finally, we show you some tools to debug any issues that you may experience.

Part IV: Making the Data Talk

In this part, we introduce the different data structures that you can use in R, such as lists and data frames. You find out how to get your data in and out of R (for example, by reading data from files or the Clipboard). You also see how to interact with other applications, such as Microsoft Excel.

Then you discover how easy it is to do some advanced data reshaping and manipulation in R. We show you how to select a subset of your data and how to sort and order it. We explain how to merge different datasets based on columns they may have in common. Finally, we show you a very powerful generic strategy of splitting and combining data and applying functions over subsets of your data. When you understand this strategy, you can use it over and over again to do sophisticated data analyses in only a few small steps.

After reading this part, you'll know how to describe and summarize your variables and data using R. You'll be able to do some classical tests (for example, calculating a t-test). And you'll know how to use random numbers to simulate some distributions.

Finally, we show you some of the basics of using linear models (for example, linear regression and analysis of variance). We also show you how to use R to predict the values of new data using models that you've fitted to your data.

Part V: Working with Graphics

They say that a picture is worth a thousand words. This is certainly the case when you want to share your results with other people. In this part, you discover how to create basic and more sophisticated plots to visualize your data. We move on from bar charts and line charts, and show you how to present cuts of your data using facets.

Part VI: The Part of Tens

In this part, we show you how to do ten things in R that you probably use Microsoft Excel for at the moment (for example, how to do the equivalent of pivot tables and lookup tables). We also give you ten tips for working with packages that are not part of base R.

Icons Used in This Book

As you read this book, you'll find little pictures in the margins. These pictures, or *icons,* mark certain types of text:

When you see the Tip icon, you can be sure to find a way to do something more easily or quickly.

You don't have to memorize this book, but the Remember icon points out some useful things that you really should remember. Usually this indicates a design pattern or idiom that you'll encounter in more than one chapter.

When you see the Warning icon, listen up. It points out something you definitely don't want to do. Although it's really unlikely that using R will cause something disastrous to happen, we use the Warning icon to alert you if something is bound to lead to confusion.

The Technical Stuff icon indicates technical information you can merrily skip over. We do our best to make this information as interesting and relevant as possible, but if you're short on time or you just want the information you absolutely *need* to know, you can move on by.

Beyond the Book

R For Dummies includes the following goodies online for easy download:

- **Cheat Sheet:** You can find the Cheat Sheet for this book here:

  ```
  www.dummies.com/cheatsheet/r
  ```

- **Extras:** We provide a few extra articles here:

  ```
  www.dummies.com/extras/r
  ```

- **Example code:** We provide the example code for the book here:

  ```
  www.dummies.com/extras/r
  ```

If we have updates to the content of the book, look here for it:

```
www.dummies.com/extras/r
```

Where to Go from Here

There's only one way to learn R: Use it! In this book, we try to make you familiar with the usage of R, but you'll have to sit down at your PC and start playing around with it yourself. Crack the book open so the pages don't flip by themselves, and start hitting the keyboard!

Part I
Getting Started with
R Programming

getting started
with

R
Programming

In this part . . .

- ✔ Introducing R programming concepts.
- ✔ Creating your first script.
- ✔ Making clear, legible code.
- ✔ Visit `www.dummies.com` for great Dummies content online.

Chapter 1

Introducing R: The Big Picture

● ●

In This Chapter

▶ Discovering the benefits of R

▶ Identifying some programming concepts that make R special

● ●

*W*ith an estimated worldwide user base of more than 2 million people, the R language has rapidly grown and extended since its origin as an academic demonstration language in the 1990s.

Some people would argue — and we think they're right — that R is much more than a statistical programming language. It's also

✔ A very powerful tool for all kinds of data processing and manipulation

✔ A community of programmers, users, academics, and practitioners

✔ A tool that makes all kinds of publication-quality graphics and data visualizations

✔ A collection of freely distributed add-on packages

✔ A versatile toolbox for extensive automation of your work

In this chapter, we fill you in on the benefits of R, as well as its unique features and quirks.

You can download R at `www.r-project.org`. This website also provides more information on R and links to the online manuals, mailing lists, conferences, and publications.

Tracing the history of R

Ross Ihaka and Robert Gentleman developed R as a free software environment for their teaching classes when they were colleagues at the University of Auckland in New Zealand. Because they were both familiar with S, a programming language for statistics, it seemed natural to use similar syntax in their own work. After Ihaka and Gentleman announced their software on the S-news mailing list, several people became interested and started to collaborate with them, notably Martin Mächler.

Currently, a group of 21 people has rights to modify the central archive of source code (`http://www.r-project.org/contributors.html`). This group is referred to as the R Core Team. In addition, many other people have contributed new code and bug fixes to the project.

Here are some milestone dates in the development of R:

- **Early 1990s:** The development of R began.

- **August 1993:** The software was announced on the S-news mailing list. Since then, a set of active R mailing lists has been created. The web page at `www.r-project.org/mail.html` provides descriptions of these lists and instructions for subscribing. (For more information, turn to "It provides an engaged community," later in this chapter.)

- **June 1995:** After some persuasive arguments by Martin Mächler (among others) to make the code available as "free software," the code was made available under the Free Software Foundation's GNU General Public License (GPL), Version 2.

- **Mid-1997:** The initial R Development Core Team was formed (although, at the time, it was simply known as the core group).

- **February 2000:** The first version of R, version 1.0.0, was released.

- **October 2004:** Release of R version 2.0.0.

- **April 2013:** Release of R version 3.0.0.

- **April 2015:** Release of R-3.2.0 (the version used in this book).

Ross Ihaka wrote a comprehensive overview of the development of R. The web page `http://cran.r-project.org/doc/html/interface98-paper/paper.html` provides a fascinating history.

Recognizing the Benefits of Using R

Of the many attractive benefits of R, a few stand out: It's actively maintained, it has good connectivity to various types of data and other systems, and it's versatile enough to solve problems in many domains. Possibly best of all, it's available for free, in more than one sense of the word.

It comes as free, open-source code

R is available under an open-source license, which means that anyone can download and modify the code. This freedom is often referred to as "free as in speech." R is also available free of charge — a second kind of freedom, sometimes referred to as "free as in beer." In practical terms, this means that you can download and use R free of charge.

As a result of this freedom, many excellent programmers have contributed improvements and fixes to the R code. For this reason, R is very stable and reliable.

Any freedom also has associated obligations. In the case of R, these obligations are described in the conditions of the license under which it is released: GNU General Public License (GPL), Version 2. The full text of the license is available at www.r-project.org/COPYING. It's important to stress that the GPL does not pertain to your usage of R. There are no obligations for using the software — the obligations just apply to redistribution. In short, if you change *and* redistribute the R source code, you have to make those changes available for anybody else to use.

It runs anywhere

The R Core Team has put a lot of effort into making R available for different types of hardware and software. This means that R is available for Windows, Unix systems (such as Linux), and the Mac.

It supports extensions

R itself is a powerful language that performs a wide variety of functions, such as data manipulation, statistical modeling, and graphics. One really big advantage of R, however, is its extensibility. Developers can easily write their own software and distribute it in the form of add-on packages. Because of the relative ease of creating and using these packages, literally thousands of packages exist. In fact, many new (and not-so-new) statistical methods are published with an R package attached.

It provides an engaged community

The R user base keeps growing. Many people who use R eventually start helping new users and advocating the use of R in their workplaces and professional circles. Sometimes they also become active on

✔ The R mailing lists (http://www.r-project.org/mail.html

✔ Question-and-answer (Q&A) websites, such as

  • StackOverflow, a programming Q&A website
    (www.stackoverflow.com/questions/tagged/r)

> • CrossValidated, a statistics Q&A website (`http://stats.stackexchange.com/questions/tagged/r`)

In addition to these mailing lists and Q&A websites, R users may

> ✔ Blog actively (`www.r-bloggers.com`).
> ✔ Participate in social networks such as Twitter (`www.twitter.com/search/rstats`).
> ✔ Attend regional and international R conferences.

See Chapter 11 for more information on R communities.

It connects with other languages

As more and more people moved to R for their analyses, they started trying to incorporate R in their previous workflows. This led to a whole set of packages for linking R to file systems, databases, and other applications. Many of these packages have since been incorporated into the base installation of R.

For example, the R package `foreign` (`http://cran.r-project.org/web/packages/foreign/index.html`) forms part of the *recommended* packages of R and enables you to read data from the statistical packages SPSS, SAS, Stata, and others (see Chapter 12).

Several add-on packages exist to connect R to database systems, such as

> ✔ RODBC, to read from databases using the Open Database Connectivity protocol (ODBC) (`http://cran.r-project.org/web/packages/RODBC/index.html`)
> ✔ ROracle, to read Oracle data bases (`http://cran.r-project.org/web/packages/ROracle/index.html`).

Initially, most of R was based on Fortran and C. Code from these two languages easily could be called from within R. As the community grew, C++, Java, Python, and other popular programming languages got more and more connected with R.

As more data analysts started using R, the developers of commercial data software no longer could ignore the new kid on the block. Many of the big commercial packages have add-ons to connect with R. Notably, both IBM's

SPSS and SAS Institute's SAS allow you to move data and graphics between the two packages, and also call R functions directly from within these packages.

Other third-party developers also have contributed to better connectivity between different data analysis tools. For example, Statconn developed RExcel, an Excel add-on that allows users to work with R from within Excel (`http://www.statconn.com/products.html`).

Looking At Some of the Unique Features of R

R is more than just a domain-specific programming language aimed at data analysis. It has some unique features that make it very powerful, the most important one arguably being the notion of *vectors*. These vectors allow you to perform sometimes complex operations on a set of values in a single command.

Performing multiple calculations with vectors

R is a vector-based language. You can think of a *vector* as a row or column of numbers or text. The list of numbers {1,2,3,4,5}, for example, could be a vector. Unlike most other programming languages, R allows you to apply functions to the whole vector in a single operation without the need for an explicit loop.

It is time to illustrate vectors with some real R code. First, assign the values 1:5 to a vector called x:

```
> x <- 1:5
> x
[1] 1 2 3 4 5
```

Next, add the value 2 to each element in the vector x:

```
> x + 2
[1] 3 4 5 6 7
```

You can also add one vector to another. To add the values `6:10` element-wise to x, you do the following:

```
> x + 6:10
[1]  7  9  11  13  15
```

To do this in most other programming language would require an explicit loop to run through each value of x. However, R is designed to perform many operations in a single step. This functionality is one of the features that make R so useful — and powerful — for data analysis.

We introduce the concept of vectors in Chapter 2 and expand on vectors and vectorization in much more depth in Chapter 4.

Processing more than just statistics

R was developed by statisticians to make statistical data analysis easier. This heritage continues, making R a very powerful tool for performing virtually any statistical computation.

As R started to expand away from its origins in statistics, many people who would describe themselves as programmers rather than statisticians have become involved with R. The result is that R is now eminently suitable for a wide variety of nonstatistical tasks, including data processing, graphical visualization, and analysis of all sorts. R is being used in the fields of finance, natural language processing, genetics, biology, and market research, to name just a few.

R is *Turing complete,* which means that you can use R alone to program anything you want. (Not every task is easy to program in R, though.)

In this book, we assume that you want to find out about R programming, not statistics, although we provide an introduction to statistics with R in Part IV.

Running code without a compiler

R is an *interpreted language,* which means that — contrary to compiled languages like C and Java — you don't need a compiler to first create a program from your code before you can use it. R interprets the code you provide directly and converts it into lower-level calls to pre-compiled code/functions.

In practice, it means that you simply write your code and send it to R, and the code runs, which makes the development cycle easy. This ease of development comes at the cost of speed of code execution, however. The downside of an interpreted language is that the code usually runs slower than the equivalent compiled code.

If you have experience in other languages, be aware that R is *not* C or Java. Although you can use R as a procedural language such as C or an object-oriented language such as Java, R is mostly based on the functional programming paradigm. As we discuss later in this book, especially in Part III, this characteristic requires a bit of a different mindset. Forget what you know about other languages, and prepare for something completely different.

Chapter 2

Exploring R

- -

In This Chapter

▶ Looking at your R editing options

▶ Starting R

▶ Writing your first R script

▶ Finding your way around the R environment

- -

*I*n order to start working in R, you need two things. First, you need a tool to easily write and edit code (an *editor*). You also need an *interface,* so you can send that code to R. Which tools you use depend to some extent on your operating system. The basic R install gives you these options:

- ✔ **Windows:** A basic user interface called **RGui.**

- ✔ **Mac OS X:** A basic user interface called **R.app.**

- ✔ **Linux:** There is no specific interface on Linux, but you can use any code editor (like Vim or Emacs) to edit your R code. R itself opens by default in a terminal window.

At a practical level, this difference between operating systems and interfaces doesn't matter very much. R is a programming language, and you can be sure that R interprets your code identically across operating systems.

Still, we want to show you how to use a standard R interface, so in this chapter we briefly illustrate how to use R with the Windows **RGui.** Our advice also works on the Mac **R.app.**

Fortunately, there is an alternative, third-party interface called *RStudio* that provides a consistent user interface regardless of operating system. RStudio increasingly is the standard editing tool for R, so we also illustrate how to use RStudio.

In this chapter, after opening an R console, you flex your R muscles and write some scripts. You do some calculations, create some numeric and text objects, take a look at the built-in help, and save your work.

Working with a Code Editor

R is many things: a programming language, a statistical processing environment, a way to solve problems, and a collection of helpful tools to make your life easier. The one thing that R is *not* is an application, which means that you have the freedom of selecting your own editing tools to interact with R.

In this section we discuss the Windows R interface, RGui (short for *R graphical user interface*). This interface also includes a very basic editor for your code. Since this standard editor is so, well, basic, we also introduce you to RStudio. RStudio offers a richer editing environment than RGui and many handy shortcuts for common tasks in R.

Alternatives to the standard R editors

Among the many freedoms that R offers you is the freedom to choose your own code editor and development environment, so you don't have to use the standard R editors or RStudio.

These are powerful full-featured editors and development environments:

- **Eclipse StatET** (www.walware.de/goto/statet): Eclipse, another powerful integrated development environment, has an R add-in called StatET. If you've done software development on large projects, you may find Eclipse useful. Eclipse requires you to install Java on your computer.

- **Emacs Speaks Statistics** (http://ess.r-project.org): Emacs, a powerful text and code editor, is widely used in the Linux world and also is available for Windows. It has a statistics add-in called Emacs Speaks Statistics (ESS), which is famous for having keyboard shortcuts for just about everything you could possibly do

and for its very loyal fan base. If you're a programmer coming from the Linux world, this editor may be a good choice for you.

- **Tinn-R** (http://nbcgib.uesc.br/lec/software/editores/tinn-r/en): This editor, developed specifically for working with R, is available only for Windows. It has some nice features for setting up collections of R scripts in projects. Tinn-R is easier to install and use than either Eclipse or Emacs, but it isn't as fully featured.

A couple of interfaces are designed as tools for special purposes:

- **Rcommander** (http://www.rcommander.com/): Rcommander provides a simple GUI for data analysis in R and contains a variety of plugins for different tasks.

- **Rattle** (http://rattle.togaware.com/): Rattle is a GUI designed for typical data mining tasks.

Exploring RGui

As part of the process of downloading and installing R, you get the standard graphical user interface (GUI), called *RGui*. RGui gives you some tools to manage your R environment — most important, a console window. The console is where you type instructions and generally get R to do useful things for you.

Seeing the naked R console

The standard installation process creates useful menu shortcuts (although this may not be true if you use Linux, because there is no standard GUI interface for Linux). In the menu system, look for a folder called R, and then find an icon called R followed by a version number (for example, R 3.2.0, as shown in Figure 2-1).

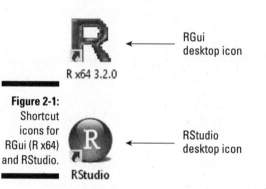

R x64 3.2.0

← RGui
desktop icon

← RStudio
desktop icon

RStudio

Figure 2-1:
Shortcut
icons for
RGui (R x64)
and RStudio.

When you open RGui for the first time, you see the R *Console* screen (shown in Figure 2-2), which lists some basic information such as your version of R and the licensing conditions.

Below all this information is the R *prompt,* denoted by a > symbol. The prompt indicates where you type your commands to R; you see a blinking cursor to the right of the prompt.

We explore the R console in more depth in "Navigating the Environment," later in this chapter.

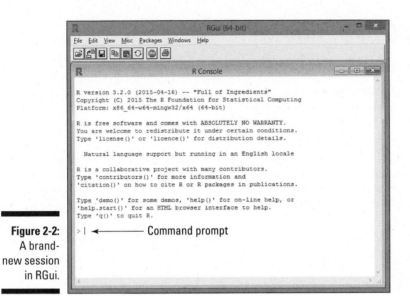

Figure 2-2:
A brand-
new session
in RGui.

Issuing a simple command

Use the console to issue a very simple command to R. Type the following to calculate the sum of some numbers, directly after the prompt:

```
> 24 + 7 + 11
```

R responds immediately to your command, calculates and displays the total in the console:

```
> 24 + 7 + 11
[1] 42
```

The answer is 42. R gives you one other piece of information: The `[1]` preceding `42` indicates that the value `42` is the first element in your answer. It is, in fact, the only element in your answer! One of the clever things about R is that it can deal with calculating many values at the same time, which is called *vector operations*. We talk about vectors later in this chapter — for now, all you need to know is that R can handle more than one value at a time.

Closing the console

To quit your R session, type the following code in the console, after the command prompt (>):

```
> quit()
```

R asks you a question to make sure that you meant to quit, as shown in Figure 2-3. Click No, because you have nothing to save. This action closes your R session (as well as RGui, if you've been using RGui as your code editor). In fact, saving a workspace image rarely is useful.

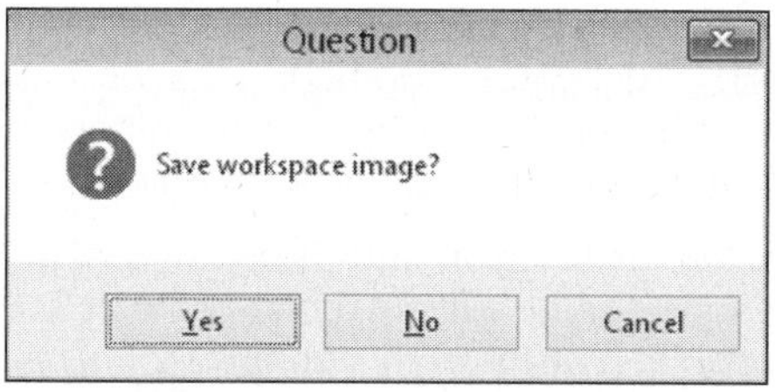

Figure 2-3:
R asks you
a simple
question.

Dressing up with RStudio

RStudio is a code editor and development environment with some very nice features that make code development in R easy and fun:

- ✔ Code highlighting that gives different colors to keywords and variables, making it easier to read

- ✔ Automatic bracket and parenthesis matching

- ✔ Code completion, so you don't have to type out all commands in full

- ✔ Easy access to R Help, with some nice features for exploring functions and parameters of functions

- ✔ Easy exploration of variables and values

Because RStudio is available free of charge for Linux, Windows, and Apple OS X, we think it's a good option to use with R. In fact, we like RStudio so much that we use it to illustrate the examples in this book. Throughout the book, you find some tips and tricks on how things can be done in RStudio. If you decide to use a different code editor, you can still use all the code examples and you'll get identical results.

To open RStudio, click the RStudio icon in your menu system or on your desktop. (You can find installation instructions in this book's appendix.)

Once RStudio starts, choose File➪New➪R Script to open a new script file.

Your screen should look like Figure 2-4. You have four work areas (also called panes):

✔ **Source:** The top-left corner of the screen contains a text editor that lets you work with source script files. Here, you can enter multiple lines of code, save your script file to disk, and perform other tasks on your script. This code editor works a bit like every other text editor you've ever seen, but it's smart. It recognizes and highlights various elements of your code, for example (using different colors for different elements), and it also helps you find matching brackets in your scripts.

✔ **Console:** In the bottom-left corner, you find the console. The console in RStudio can be used in the same way as the console in RGui (refer to "Seeing the naked R console," earlier in this chapter). This is where you do all the interactive work with R.

✔ **Environment and History:** The top-right corner is a handy overview of your environment, where you can inspect the variables you created in your session, as well as their values. (We discuss the environment in more detail later in this chapter.) This is also the area where you can see a history of the commands you've issued in R.

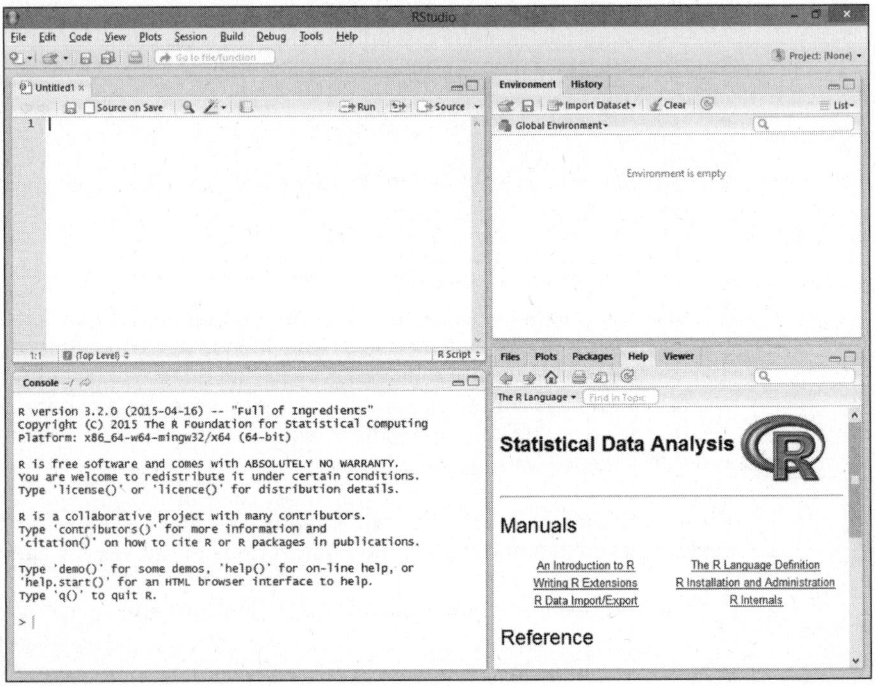

Figure 2-4: RStudio's four work areas (panes).

✔ **Files, plots, package, help, and viewer:** In the bottom-right corner, you have access to several tools:

- **Files:** This is where you can browse the folders and files on your computer.

- **Plots:** This is where R displays your plots (charts or graphs). We discuss plots in Part V.

- **Packages:** You can view a list of all installed packages.

A *package* is a self-contained set of code that adds functionality to R, similar to the way that add-ins add functionality to Microsoft Excel.

- **Help:** This is where you can browse R's built-in Help system.

- **Viewer:** This is where RStudio displays previews of some advanced features, such as dynamic web pages and presentations that you can create with R and add-on packages.

Starting Your First R Session

By now, you probably are itching to get started on some real code. In this section, you get to do exactly that. Get ready to get your hands dirty!

Saying hello to the world

Programming books typically start with a very simple program. Often, this first program creates the message `"Hello world!"`. In R, hello world program consists of one line of code.

Start a new R session, type the following in your console, and press Enter:

```
> print("Hello world!")
```

R responds immediately with this output:

```
[1] "Hello world!"
```

As we explain in the introduction to this book, we collapse input and output into a single block of code, like this:

```
> print("Hello world!")
[1] "Hello world!"
```

Doing simple math

Type the following in your console to calculate the sum of five numbers:

```
> 1 + 2 + 3 + 4 + 5
[1] 15
```

The answer is 15, which you can easily verify for yourself. You may think that there's an easier way to calculate this value, though — and you'd be right. We explain how in the following section.

Using vectors

A *vector* is the simplest type of data structure in R. The R manual defines a vector as "*a single entity consisting of a collection of things*". A collection of numbers, for example, is a numeric vector — the first five integer numbers form a numeric vector of length 5.

To construct a vector, type into the console:

```
> c(1, 2, 3, 4, 5)
[1] 1 2 3 4 5
```

In constructing your vector, you have successfully used a function in R. In programming language, a *function* is a piece of code that takes some inputs and does something specific with them. In constructing a vector, you tell the c() function to construct a vector with the first five integers. The entries inside the parentheses are referred to as *arguments*.

You also can construct a vector by using operators. An *operator* is a symbol you stick between two values to make a calculation. The symbols +, -, *, and / are all operators, and they have the same meaning they do in mathematics. Thus, 1+2 in R returns the value 3, just as you'd expect.

One very handy operator is called *sequence,* and it looks like a colon (:). Type the following in your console:

```
> 1:5
[1] 1 2 3 4 5
```

That's more like it. With three keystrokes, you've generated a vector with the values 1 through 5. To calculate the sum of this vector, type into your console:

```
> sum(1:5)
[1] 15
```

While quite basic, this example shows you that using vectors allows you to do complex operations with a small amount of code. As vectors are the smallest possible unit of data in R, you get to work with vectors extensively in later chapters.

Storing and calculating values

Using R as a calculator is very interesting but perhaps not all that useful. A much more useful capability is storing values and then doing calculations on these stored values. Try this:

```
> x <- 1:5
> x
[1]  1 2 3 4 5
```

In these two lines of code, you first assign the sequence 1:5 to an object called x. Then you ask R to print the value of x by typing x in the console and pressing Enter.

In R, the assignment operator is <-, which you type in the console by using two keystrokes: the less-than symbol (<) followed by a hyphen (-). The combination of these two symbols represents assignment. It's good practice to always surround the <- with spaces. This makes your code much easier to read and understand.

In addition to retrieving the value of a variable, you can do calculations on that value. Create a second variable called y, and assign it the value 10. Then add the values of x and y, as follows:

```
> y <- 10
> x + y
[1]  11 12 13 14 15
```

The values of the two variables themselves don't change unless you assign a new value to either of them. You can check this by typing the following:

```
> x
[1]  1 2 3 4 5
> y
[1]  10
```

Now create a new variable z, assign it the value of x + y, and print its value:

```
> z <- x + y
> z
[1]  11 12 13 14 15
```

Variables also can take on text values. You can assign the value `"Hello"` to a variable called h, for example, by presenting the text to R inside quotation marks, like this:

```
> h <- "Hello"
> h
[1] "Hello"
```

You must enter text or character values to R inside quotation marks — either single or double. R accepts both. So both `h <- "Hello"` and `h <- 'Hello'` are examples of valid R syntax. For consistency, we use double quotes throughout this book.

In "Using vectors," earlier in this chapter, you use the `c()` function to combine numeric values into vectors. This technique also works for text:

```
> hw <- c("Hello", "world!")
> hw
[1] "Hello" "world!"
```

You use the `paste()` function to concatenate multiple text elements. By default, `paste()` puts a space between the different elements, like this:

```
> paste("Hello", "world!")
[1] "Hello world!"
```

Talking back to the user

You can write R scripts that have some interaction with a user. To ask the user questions, you can use the `readline()` function. In the following code snippet, you read a value from the keyboard and assign it to the variable yourname:

```
> h <- "Hello"
> yourname <- readline("What is your name? ")
What is your name? Andrie
> paste(h, yourname)
[1] "Hello Andrie"
```

This code seems to be a bit cumbersome, however. Clearly, it would be much better to send these three lines of code simultaneously to R and get them evaluated in one go. In the next section, we show you how.

Sourcing a Script

Until now, you've worked directly in the R console and issued individual commands in an *interactive* style of coding. In other words, you issue a command, R responds, you issue the next command, R responds, and so on.

In this section, you kick it up a notch and tell R to perform several commands one after the other without waiting for additional instructions. Because the R function to run an entire script is source(), R users refer to this process as *sourcing a script.*

To prepare your script to be sourced, you first write the entire script in an editor window. In RStudio, for example, the editor window is in the top-left corner of the screen (refer to Figure 2-4). Whenever you press Enter in the editor window, the cursor moves to the next line, as in any text editor.

To create a new script in RStudio, begin by opening the editor window (choose File ⇨ New File ⇨ R script to open the editor window). Type the following lines of code in the editor window. Notice that the last line contains a small addition to the code you saw earlier: the print() function.

```
h <- "Hello"
yourname <- readline("What is your name?")
print(paste(h, yourname))
```

Remember to type the print() function as part of your script. Sourced scripts behave differently from interactive code in printing results. In interactive mode, a result is printed without needing to use a print() function. But when you source a script, output is by default printed only if you have an explicit print() function.

You can type multiple lines of code into the source editor without having each line evaluated by R. Then, when you're ready, you can send the instructions to R — in other words, source the script.

When you use RGui or RStudio, you can do this in one of three ways:

- ✔ **Send an individual line of code from the editor to the console.** Click the line of code you want to run, and then press Ctrl+R in RGui. In RStudio, you can press Ctrl+Enter or click the Run button.

- ✔ **Send a block of highlighted code to the console.** Select the block of code you want to run, and then press Ctrl+R (in RGui) or Ctrl+Enter (in RStudio).

- ✔ **Send the entire script to the console (which is called sourcing a script).** In RGui, click anywhere in your script window, and then choose Edit ⇨ Run all. In RStudio, click anywhere in the source editor, and press Ctrl+Shift+S or click the Source button.

These keyboard shortcuts are defined only in RGui or RStudio. If you use a different source editor, you may have different options.

Now you can send the entire script to the R console. To do this, click the Source button in the top-right corner of the editor window or choose Edit➪Source. The script starts, reaches the point where it asks for input, and then waits for you to enter your name in the console window. Your screen should now look like Figure 2-5. Notice that the *Environment* pane now lists the two objects you created: h and yourname.

When you click the Source button, source('~/.active-rstudio-document') appears in the console. What RStudio actually does here is save your script in a temporary file and then use the R function source() to call that script in the console. Remember this function; you'll meet it again.

Echoing your work

If you click on the little arrow next to the Source button in RStudio, you see two different source options, as shown in Figure 2-6. By clicking the Source button before, you used the option without echo. This means that R will run the complete script at once, but won't send any output to the console.

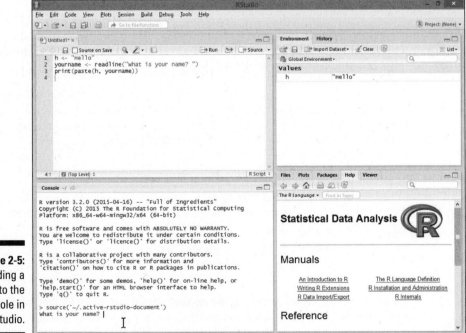

Figure 2-5: Sending a script to the console in RStudio.

Figure 2-6:
Sourcing
your code
with or
without
echo in
RStudio

If you click on the second option, R runs again the complete script in one go, but this time it will show every individual line in the console. So both options differ only in the output you see. You can safely try out both options to compare.

You can use the echo option also outside RStudio by using the source() function with the argument echo set to TRUE. We explain functions and arguments in Chapter 3, and far more detailed again in Chapter 8.

Whether you source with or without echo doesn't make any difference regarding the results of your code. You can use the echo option if you want to source a long script and keep track of which part of the script R is currently carrying out.

Finding help on functions

We discuss R's built-in help system in Chapter 11, but for now, to get help on any function, type ? in the console. To get help with the paste() function, for example, type the following:

```
> ?paste
```

This code opens a Help window. In RStudio, this Help window is in the bottom-right corner of your screen by default. In other editors, the Help window sometimes appears as a local web page in your default web browser.

You also can type help, but remember to use parentheses around your search term:

```
> help(paste)
```

Navigating the Environment

So far in this chapter, you've created several variables. These form part of what R calls the global environment. The global environment refers to all the variables and functions (collectively called *objects*) that you create during the session, as well as any packages that are loaded.

Often, you want to remind yourself of all the variables you've created in the environment. To do this, use the `ls()` function to list the objects in the environment. In the console, type the following:

```
> ls()
[1] "h"        "hw"       "x"        "y"        "yourname" "z"
```

R tells you the names of all the variables that you created.

One very nice feature of RStudio lets you examine the contents of the environment at any time without typing any R commands. By default, the top-right window in RStudio has two tabs: *Environment* and *History*. Click the Environment tab to see the variables in your global environment, as well as their values. For example, in Figure 2-5 you see that the global environment contains one object called h that contains the value `"hello"`.

Manipulating the content of the environment

If you decide that you don't need some variables anymore, you can remove them. Suppose that the object z is simply the sum of two other variables and no longer needed. To remove it permanently, use the `rm()` function and then use the `ls()` function to display the contents of the environment, as follows:

```
> rm(z)
> ls()
[1] "h"        "hw"       "x"        "y"        "yourname"
```

Notice that the object z is no longer there.

Saving your work

You have several options for saving your work:

- You can save individual variables with the `save()` function.
- You can save the entire environment with the `save.image()` function.
- You can save your R script file, using the appropriate save menu command in your code editor.

Suppose you want to save the value of `yourname`. To do that, follow these steps:

1. **Find out which working directory R will use to save your file by typing the following:**

```
> getwd()
[1] "c:/users/andrie"
```

The default working directory should be your user folder. The exact name and path of this folder depend on your operating system. (In Chapter 12, you get more familiar with the working directory.)

If you use the Windows operating system, the path is displayed with slashes instead of backslashes. In R, similar to many other programming languages, the backslash character has a special meaning. The backslash indicates an *escape sequence,* indicating that the character following the backslash means something special. For example, \t indicates a tab, rather than the letter *t.* (You can read more about escape sequences in Chapter 12.) Rest assured that, although the working directory is displayed differently from what you're used to, R is smart enough to translate it when you save or load files. Conversely, when you type a file path, you have to use slashes, not backslashes.

2. **Type the following code in your console, using a filename like `yourname.rda`, and then press Enter.**

```
> save(yourname, file = "yourname.rda")
```

R silently saves the file in the working directory. If the operation is successful, you don't get any confirmation message.

3. **To make sure that the operation was successful, use your file browser to navigate to the working directory, and see whether the new file is there.**

You have a file browser in the lower-right panel of RStudio.

Retrieving your work

To retrieve saved data, you use the load() function. Say you want to retrieve the value of yourname that you saved previously.

First, remove the variable yourname, so you can see the effect of the load process:

```
> rm(yourname)
```

If you're using RStudio, you may notice that yourname is no longer displayed in the Environment pane.

Next, use load to retrieve your variable. Type **load** followed by the filename you used to save the value earlier:

```
> load("yourname.rda")
```

Notice that yourname reappears in the Environment pane of RStudio.

Chapter 3

The Fundamentals of R

*B*efore you start discovering the different ways you can use R on your data, you need to know a few more fundamental things about R.

In Chapter 2, we show you how to use the command line and work with the global environment, so if you read that chapter, you can write a simple script and use the print(), paste(), and readline() functions — at least in the most basic way. But functions in R are more complex than that, so in this chapter we tell you how to get the most out of your functions.

As you add more arguments to your functions and more functions to your scripts, those scripts can become pretty complex. To keep your code clear — and yourself sane — you can follow the basic organizational principles we cover in this chapter.

Finally, much of R allows you to use other people's code very easily. You can extend R with packages that have been contributed to the R community by hundreds of developers. In this chapter, we tell you where you can find these packages and how you can use them in R.

Using the Full Power of Functions

Functions form the core of R; everything you do in R uses a function in one way or another. More importantly, the way functions work in R allows you to carry out multiple complex operations in one step or a few simple steps. In this section, we show you how you can use functions the smart way. First, you learn about the key property of functions that makes R so different from other programming languages. Then we tell you how you can reach a whole set of functionalities in R functions with arguments. Finally, we tell you how you can save the history of all the commands you've used in a session with — you guessed it! — a function.

Vectorizing your functions

Vectorized functions are a very useful feature of R, but programmers who are used to other languages often have trouble with this concept at first. A *vectorized* function works not just on a single value, but on a whole vector of values at the same time. Your natural reflex as a programmer may be to loop over all values of the vector and apply the function on every element, but vectorization makes that unnecessary. Trust us: When you start using vectorization in R, it'll help simplify your code.

To try vectorized functions, you have to make a vector. You do this by using the c() function, which stands for *combine*. The actual values are separated by commas.

Here's an example: Suppose that Granny plays basketball with her friend Geraldine, and you keep a score of Granny's number of baskets in each game. After six games, you want to know how many baskets Granny has made so far this season. You can combine these numbers into a vector, like this:

```
> baskets.of.Granny <- c(12, 4, 4, 6, 9, 3)
> baskets.of.Granny
[1] 12 4 4 6 9 3
```

To find the total number of baskets Granny made, you just type the following:

```
> sum(baskets.of.Granny)
[1] 38
```

You could get the same result by going over the vector number by number, adding each new number to the sum of the previous numbers. But that method would require you to write more code and it would take longer to calculate. You won't notice it on just six numbers, but the difference will be obvious when you have to sum a few thousand of them.

Actually, this kind of vectorization occurs in many programming languages. Functions that work this way summarize the data in a vector; they take all values in the vector and calculate a single result.

R also can carry out functions along vectors. This type of vectorization is pretty unique, and forms the core of R's incredible power. Quite a few people have difficulties grasping that behavior in the beginning, but it's easy to understand when you see it happen.

To see how it works, try using the `paste()` function. First, you construct two vectors (for example, a vector with first names and a vector with last names). To create a vector with the full names from the original vectors, you can simply use the `paste()` function, like this:

```
> firstnames <- c("Andrie", "Joris")
> lastnames <- c("de Vries", "Meys")
> paste(firstnames, lastnames)
[1] "Andrie de Vries" "Joris Meys"
```

R automatically loops over the values of each vector, and concatenates (pastes) them together, element by element. So the first value of the vector `firstnames` is pasted to the first value of `lastnames`, the second value of `firstnames` to the second of `lastnames`, and so forth. That's how vectorization works.

What happens if both vectors don't have the same amount of values? If you make a vector with the first names of the members of your family, `paste()` can add the last name to all of them with one command, as in the following example:

```
> firstnames <- c("Joris", "Carolien", "Koen")
> lastname <- "Meys"
> paste(firstnames,lastname)
[1] "Joris Meys" "Carolien Meys" "Koen Meys"
```

R takes the vector `firstnames` and then pastes the `lastname` into each value. How cool is that? Actually, R again combines two vectors. The second vector — in this case, `lastname` — is only one value long. That value gets *recycled* by the `paste()` function as long as necessary (for more on recycling, turn to Chapter 4).

So to process multiple values in R, you don't need complicated code. All you have to do is make the vectors and put them in the function. In Chapter 5, you can find more information about the power of `paste()`.

Putting the argument in a function

Most functions in R have arguments that allow you to specify exactly what you want the function to do. All these arguments also have a name. For example, the first argument of the `print()` function is called `x`. You can check this yourself by looking at the help file of the function using `?print`.

By specifying an argument, in other words passing a value to that argument, you tell the function what you want to do. So if you use `print("Hello world!")`, you actually pass the value `"Hello world!"` to the argument x of the `print()` function. The `print()` function tells R that you want to print something, and the value for the argument x tells R what exactly you want to print.

In R, you have two general types of arguments:

✔ Arguments with default values
✔ Arguments without default values

If an argument has no default value, the value may be optional or required. In general, the first argument is almost always required. Try entering the following:

```
> print()
```

R tells you that it needs the argument x specified:

```
Error in .Internal(print.default(x, digits, quote, na.print, print.gap, : 'x'
                is missing
```

You can pass a value to an argument using the = sign like this:

```
> print(x = "Isn't this fun?")
```

Sure it is. But wait — when you entered the `print("Hello world!")` command in Chapter 2, you didn't add the name of the argument, and the function worked. That's because R knows the names of the arguments and just assumes that you pass them in exactly the same order as they're shown in the usage line of the Help page for that function. (For more information on reading the Help pages, turn to Chapter 11.)

If you type the values for the arguments in Help-page order, you don't have to specify the argument names. You can list the arguments in any order you want, as long as you specify their names.

Try entering the following example:

```
> print(digits = 4, x = 11/7)
[1] 1.571
```

You may wonder where the `digits` argument comes from, because it's not explained in the Help page for `print()`. That's because it isn't an argument of the `print()` function itself, but of the function `print.default()`. Take a look again at the error you got if you typed **print()**. R mentions the `print.default()` function instead of the `print()` function.

In fact, `print()` is called a *generic function*. It determines the type of the object that's passed as an argument and then looks for a function that can deal with this type of object. That function is called the *method* for the specific object type. In case there is no specific function, R calls the *default method*. This is the function that works on all object types that have no specific method. In this case, that's the `print.default()` function. Keep in mind that a default method doesn't always exist. We explain this in more detail in Chapter 8. For now, just remember that arguments for a function can be shown on the Help pages of different methods.

If you forget which arguments to use, you can find that information in the Help files. Don't forget to look at the arguments of specific methods as well. You often find a link to those specific methods at the bottom of the Help page. For example, to read the help for the `paste()` function, type `?paste` into your R console.

Making history

By default, R keeps track of all the commands you use in a session. This tracking can come in handy if you need to reuse a command you used earlier or want to keep track of the work you did before. These previously used commands are kept in the history.

You can browse the history from the command line by pressing the up-arrow and down-arrow keys. When you press the up-arrow key, you get the commands you typed earlier at the command line. You can press Enter at any time to run the command that is currently displayed.

Saving the history is done using the `savehistory()` function. By default, R saves the history in a file called `.Rhistory` in your current working directory. This file is automatically loaded again the next time you start R, so you have the history of your previous session available.

If you want to use another filename, use the argument `file` like this:

```
> savehistory(file = "Chapter3.Rhistory")
```

Be sure to add the quotation marks around the filename.

You can open a file explorer window and take a look at the history by opening the file in any text editor, like Notepad.

You don't need to use the file extension `.Rhistory` — R doesn't care about extensions that much. But using `.Rhistory` as a file extension will make it easier to recognize as a history file.

If you want to load a history file you saved earlier, you can use the `load-history()` function. This will replace the history with the one saved in the `.Rhistory` file in the current working directory. If you want to load the history from a specific file, you use the `file` argument again, like this:

```
> loadhistory("Chapter3.Rhistory")
```

Keeping Your Code Readable

You may wonder why you should bother about reading code. You wrote the code yourself, so you should know what it does, right? You do now, but will you remember what you did if you have to redo that analysis six months from now on new data? Besides, you may have to share your scripts with other people, and what seems obvious to you may be far less obvious for them.

Some of the rules you're about to see aren't that strict. In fact, you can get away with almost anything in R, but that doesn't mean it's a good idea. In this section, we explain why you should avoid some constructs even though they aren't strictly wrong.

Following naming conventions

R is very liberal when it comes to names for objects and functions. This freedom is a great blessing and a great burden at the same time. Nobody is obliged to follow strict rules, so everybody who programs something in R can basically do as he or she pleases.

Choosing a correct name

Although almost anything is allowed when giving names to objects, there are still a few rules in R that you can't ignore:

✔ **Names must start with a letter or a dot.** If you start a name with a dot, the second character can't be a digit.

✔ **Names should contain only letters, numbers, underscore characters (_), and dots (.).** Although you can force R to accept other characters in names, you shouldn't, because these characters often have a special meaning in R.

✔ **You can't use the following special keywords as names:**

- `break`

- `else`

- FALSE
- for
- function
- if
- Inf
- NA
- NaN
- next
- NULL
- repeat
- return
- TRUE
- while

R is *case sensitive,* which means that, for R, `lastname` and `Lastname` are two different objects. If R tells you it can't find an object or function and you're sure it should be there, check to make sure you used the right case.

Choosing a clear name

When you start writing code, it's tempting to use short, generic names like `x`. There's nothing wrong with that, as long as it is clear what each object represents. But that might become difficult when all your objects have a single letter name. Likewise, calling your datasets `data1`, `data2,` and so forth may be a bit confusing for the person who has to read your code later on, even if it makes all kinds of sense to you now. ***Remember:*** You could be the one who, in three months, is trying to figure out exactly what you were trying to achieve. Using descriptive names will allow you to keep your code readable.

Although you can name an object almost whatever you want, some names will cause less trouble than others. You may have noticed that none of the functions we've used until now are mentioned as being off-limits (see the preceding section). That's right: If you want to call an object `paste`, you're free to do so:

```
> paste <- paste("This gets","confusing")
> paste
[1] "This gets confusing"
> paste("Don't","you","think?")
[1] "Don't you think?"
```

R almost always will know perfectly well when you want the vector `paste` and when you need the function `paste()`. That doesn't mean it's a good idea to use the same name for both items, though. In some cases, doing so can cause unexpected errors. So if you can avoid giving the name of a function to an object, you should.

One situation in which you can really get into trouble is when you use capital `F` or `T` as an object name. You can do it, but you're likely to break code at some point. Although it's a very bad idea, `T` and `F` are all too often used as abbreviations for `TRUE` and `FALSE`, respectively. But `T` and `F` are not reserved keywords. So, if you change them, R will first look for the object `T` and only then try to replace `T` with `TRUE`. And any code that still expects `T` to mean `TRUE` will fail from this point on. Never use `F` or `T`, not as an object name and not as an abbreviation.

Choosing a naming style

If you have experience in programming, you've probably heard of *camel case* before. Camel case is a way of giving longer names to objects and functions. You capitalize every first letter of a word that is part of the name to improve the readability. So, you can have a `veryLongVariableName` and still be able to read it.

Unlike many other languages, R doesn't use the dot (`.`) as an operator, so the dot can be used in names for objects as well. This style is called *dotted style,* where you write everything in lowercase and separate words or terms in a name with a dot. In fact, in R, many function names use dotted style. You've met a function like this earlier in the chapter: `print.default()`. Some package authors also use an underscore instead of a dot.

`print.default()` is the default method for the `print()` function. You can find Information on the arguments of the Help page for `print.default()`.

You're not obligated to use dotted style; you can use whatever style you want. We use dotted style throughout this book for objects, and camel case for functions. R uses dotted style for many base functions and objects, but because some parts of the internal mechanisms of R rely on that dot, you're safer using camel case for functions. Whenever you see a dot, though, you don't have to wonder what it does — it's just part of the name.

The whole naming issue reveals one of the downsides of using open-source software: It's written by very intelligent and unselfish people with very strong opinions, so the naming of functions in R is far from standardized.

Consistent inconsistency

You would expect the function to be called `save.history()`, but it's called `savehistory()` without the dot. Likewise, you might expect a function `R.version()`, but instead it's `R.Version()`. (`R.Version()` gives you all the information on the version of R you're running, including the platform you're running it on.) Sometimes, the people writing R use camel case: If you want to get only the version number of R, you have to use the function `getRversion()`. Some package authors choose to use underscores (_) instead of dots for separation of the words; this style is used often within some packages we discuss later in this book (for example, the `ggplot2` package in Chapter 18).

Structuring your code

Names aren't the only things that can influence the readability of your code. When you start nesting functions or perform complex calculations, your code can turn into a big mess of text and symbols rather quickly. Luckily, you have some tricks to clear up your code so you can still decipher what you did three months down the road.

Nesting functions and doing complex calculations can lead to very long lines of code. If you want to make a vector with the names of your three most beloved song titles, for example, you're already in for trouble. Luckily, R lets you break a line of code over multiple lines in your script, so you don't have to scroll to the right the whole time.

You don't even have to use a special notation or character. R will know that the line isn't finished as long as you give it some hint. Generally speaking, you have to make sure the command is undoubtedly incomplete. There are several ways to do that:

- ✔ **You can use a quotation mark to start a string.** R will take all the following input — including the line breaks — as part of the string, until it meets the matching second quotation mark.

- ✔ **You can end the incomplete line with an operator (like +, /, <-, and so on).** R will know that something else must follow. This lets you create structure in longer calculations.

- ✔ **You can open a parenthesis for a function.** R will read all the input it gets as one line until it meets the matching parenthesis. This allows you to line up arguments below a function, for example.

The following little script shows all these techniques:

```
baskets.of.Geraldine <-
        c(5, 3, 2, 2, 12, 9)

Intro <- "It is amazing! The All Star Grannies scored
a total of"
Outro <- "baskets in the last six games!"

Total.baskets <- baskets.of.Granny +
                baskets.of.Geraldine

Text <- paste(Intro,
              sum(Total.baskets),
              Outro)
cat(Text)
```

You can copy this code into a script file and run it in the console. If you run this little snippet of code, you see the following output in the console:

```
It is amazing! The All Star Grannies scored
a total of 71 baskets in the last six games!
```

This immediately shows what the cat() function does. It prints whatever you pass as an argument directly to the console. It also interprets special characters like line breaks and tabs. If you look at the vector Text, you will see this:

```
> Text
[1] "It is amazing! The All Star Grannies scored \na total of 71 baskets in the
            last six games!"
```

The \n represents the line break. Even though it's pasted to the a, R will recognize \n as a separate character. (You can find more information on special characters in Chapter 12.)

All this also works at the command line. If you type an unfinished command, R will change the prompt to a + sign, indicating that you can continue to type your command:

```
> cat("If you doubt whether it works,
+ just try it out.")
If you doubt whether it works,
just try it out.
```

RStudio automatically adds a line break at the end of a cat() statement if there is none, but R doesn't do that. So, if you don't use RStudio, remember to add a line break (or the symbol \n) at the end of your string.

Adding comments

Often, you want to add a bit of extra information to a script file. You may want to tell who wrote it and when. You may want to explain what the code does and what all the variable names mean.

You can do this by typing that information after the # symbol (usually called *hash* or *pound*). R ignores everything that appears after the hash symbol. You can use the hash symbol at the beginning of a line or somewhere in the middle. Run the following script, and see what happens:

```
# The All Star Grannies do it again!
baskets.of.Granny <- c(12,4,4,6,9,3) # Granny rules
sum(baskets.of.Granny) # total number of baskets
```

R has no specific construct to spread a comment over multiple lines. You'll have to precede every line of the comment block with a hash symbol (#). In RStudio, you can easily comment or uncomment several lines together by selecting them and pressing Ctrl+Shift+C. Other editors often have similar shortcuts.

Getting from Base R to More

Until now, you've used only functions that are available in the basic installation of R. But the real power of R lies in the fact that anyone can write their own functions and share them with other R users in an organized manner. Many knowledgeable people have written convenient functions with R, and often a new statistical method is published together with R code. Most of these authors distribute their code as R *packages* (collections of R code, Help files, datasets, and so on that can be incorporated easily into R itself). In this section, we tell you how to find and add packages to your R installation.

Finding packages

Several websites, called *repositories,* offer a collection of R packages. The most important repository is the Comprehensive R Archive Network (CRAN; http://cran.r-project.org), which you can access easily from within R.

In addition to housing the installation files for R itself (see Appendix A) and a set of manuals for R, CRAN contains a collection of package files and the reference manuals for all packages. For some packages, a *vignette* (which gives you a short introduction to the use of the functions in the package) is also available. Finally, CRAN lets you check whether a package is still maintained and see an overview of the changes made in the package. CRAN is definitely worth checking out!

Installing packages

You install a package in R with the function — wait for it — install. packages(). Who could've guessed? So, to install the fortunes package, for example, you simply pass the name of the package as a string to the install.packages() function.

The fortunes package contains a whole set of humorous and thought-provoking quotes from mailing lists and help sites. You install the package like this:

```
> install.packages("fortunes")
```

R may ask you to specify a CRAN mirror. Because everyone in the whole world has to access the same servers, CRAN is mirrored on more than 80 registered servers, often located at universities. Pick one that's close to your location, and R will connect to that server to download the package files. In RStudio, you can set the mirror by choosing Tools⇨Global Options⇨Packages.

Next, R gives you some information on the installation of the package:

```
Installing package(s) into 'D:/R/library'(as 'lib' is unspecified)
....
opened URL
downloaded 165 Kb

package 'fortunes' successfully unpacked and MD5 sums checked
....
```

It tells you which directory (called a *library*) the package files are installed in, and it tells you whether the package was installed successfully. Granted, it does so in a rather technical way, but the word *successfully* tells you everything is okay.

Loading and unloading packages

After a while, you can end up with a collection of many packages. If R loaded all of them at the beginning of each session, that would take a lot of memory and time. So, before you can use a package, you have to load it into R by using the library() function.

You load the `fortunes` package like this:

```
> library("fortunes")
```

You don't have to put quotation marks around the package name when using `library()`, but it is wise to do so.

Now you can use the functions from this package at the command line, like this:

```
> fortune("This is R")
```

The *library* is the directory where the packages are installed. Never, ever call a package a library. That's a mortal sin in the R community. Take a look at the following, and never forget it again:

```
> fortune(161)
```

You can use the `fortune()` function without arguments to get a random selection of the `fortunes` available in the package. It's a nice read.

If you want to unload a package, you'll have to use some R magic. The `detach()` function will let you do this, but you have to specify that it's a package you're detaching and that you want to unload it, like this:

```
> detach(package:fortunes, unload=TRUE)
```

Actually, even this line of code doesn't always unload a package. For example, if a package is used by another package that's still loaded, that code won't work. If you've been toying around in R for a while and tried to load and unload many packages, save your work, close R, and start a fresh session.

A package is as good as its author

Many people contribute in one way or another to R. As in any open-source community, there are people with very strong coding skills and people with especially heartwarming enthusiasm. R itself is tested thoroughly, and packages available on CRAN are checked for safety and functionality. This means that you can safely download and use those packages without fear of breaking your R installation or — even worse — your computer.

It doesn't mean, however, that the packages always do what they claim to do. After all, even programmers are human, and they make mistakes. Before you use a new package, you may want to test it out using an example where you know what the outcome should be.

(continued)

(continued)

But the fact that many people use R is an advantage. Whereas reporting errors to a huge company can be a proverbial pain in the lower regions, you can reach the authors of packages by email. Users report bugs and errors all the time, and package authors continue to update and improve their packages. Overall, the quality of the packages is at least as good as, if not better than, the quality of commercial applications. After all, the source code of every package is readily available for anyone to check and correct. In this way, both R and the packages improve with contributions from users.

So, in fact, we could have titled this sidebar "A package is as good as the community that uses it." The R community is simply fantastic — and you should know. By buying this book, you officially became a member.

Part II
Getting Down to Work in R

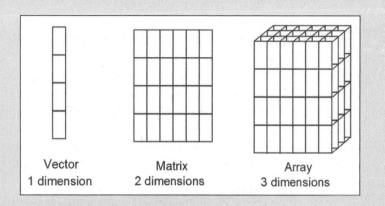

Vector
1 dimension

Matrix
2 dimensions

Array
3 dimensions

In this part . . .

- ✔ Applying the building blocks.
- ✔ Representing your data.
- ✔ Applying important operators.
- ✔ Visit www.dummies.com/extras/r for great Dummies content online.

Chapter 4

Getting Started with Arithmetic

In This Chapter
▶ Using R as a fancy calculator
▶ Constructing and working with vectors
▶ Vectorizing your calculations

*S*tatistics isn't called *applied mathematics* just for the fun of it. Every statistical analysis involves a lot of calculations, and calculation is what R is designed for — the work that R does best.

R goes far beyond employing the classic arithmetic operators. It also contains sets of operators and functions that work on complete vectors at the same time. If you've read Chapter 3, you've gotten a first glimpse into the power of vectorized functions. In this chapter, you discover the full power of vectorization, using it to speed up calculations and perform complex tasks with very little code.

We can't possibly cover all of R's mathematical functions in one chapter — or even in one book — so we encourage you to browse the Help files and other sources when you're done with this chapter. You can find search tips and a list of interesting sources in Chapter 11.

Working with Numbers, Infinity, and Missing Values

In many low-level computer languages, numerical operators are limited to performing standard arithmetic operations and some convenient functions like sum() and sqrt(). For R, the story is a bit different. R has four different groups of mathematical operators and functions:

✔ **Basic arithmetic operators:** These operators are used in just about every programming language. We discuss some of them in Chapter 2.

✔ **Mathematical functions:** You can find these advanced functions on a technical calculator.

✔ **Vector operations:** Vector operations are functions that make calculations on a complete vector, like sum(). Each result depends on more than one value of the vector.

✔ **Matrix operations:** These functions are used for operations and calculations on matrices.

In the following sections, you get familiar with basic arithmetic operators, mathematical functions, and vector operations. We cover matrix operations in Chapter 7.

Doing basic arithmetic

R has a rather complete set of arithmetic operators, so you can use R as a fancy calculator, as you see in this section.

Using arithmetic operators

Table 4-1 lists some basic arithmetic operators. Most of them are very familiar to programmers (and anybody else who studied math in school).

All these operators are vectorized. Chapter 3 shows the use of a vectorized function with the paste() function, and the process works exactly the same way with operators. By using vectorized operators, you can carry out complex calculations with minimal code.

Table 4-1	Basic Arithmetic Operators	
Operator	*Description*	*Example*
x + y	y added to x	2 + 3 = 5
x − y	y subtracted from x	8 − 2 = 6
x * y	x multiplied by y	3 * 2 = 6
x / y	x divided by y	10 / 5 = 2
x ^ y	x raised to the power y	2 ^ 5 = 32
x %% y	remainder of x divided by y (x mod y)	7 %% 3 = 1
x %/% y	x divided by y but rounded down (integer divide)	7 %/% 3 = 2

To see how this works, consider these two vectors, which we first discuss in the All-Star Grannies example in Chapter 3. One vector represents the number of baskets Granny made during the six games of the basketball season, and the other one represents the number of baskets her friend Geraldine made:

```
> baskets.of.Granny <- c(12, 4, 4, 6, 9, 3)
> baskets.of.Geraldine <- c(5, 3, 2, 2, 12, 9)
```

Suppose that Granny and Geraldine decide to raise money for the Make-A-Wish Foundation and asked people to make a donation for every basket they made. Granny requested $120 per basket, and Geraldine asked for $145 per basket. How do you calculate the total donations that they collected for each game?

R makes the calculation easy. First, calculate how much each lady earned per game, as follows:

```
> Granny.money <- baskets.of.Granny * 120
> Geraldine.money <- baskets.of.Geraldine * 145
```

In this example, every value in the vector is multiplied by the amount of money. Check for yourself by taking a look at the values in Granny.money and Geraldine.money.

To get the total money these ladies earned in each game, you simply do this:

```
> Granny.money + Geraldine.money
[1] 2165 915 770 1010 2820 1665
```

You also could do this whole calculation in a single line, as follows:

```
> baskets.of.Granny * 120 + baskets.of.Geraldine * 145
[1] 2165 915 770 1010 2820 1665
```

Controlling the order of the operations

In the previous example, you used both a multiplication and an addition operator. As you see from the result, R correctly multiplies all numbers before adding them together. For all arithmetic operators, the classic rules for the order of operations apply. Calculations are carried out in the following order:

1. **Exponentiation**

2. **Multiplication and division in the order in which the operators are presented**

3. **Addition and subtraction in the order in which the operators are presented**

The mod operator (%%) and the integer division operator (%/%) have the same priority as the normal division operator (/) in calculations.

You can change the order of the operations by using parentheses, like this:

```
>4 + 2 * 3
[1] 10
> (4 + 2)* 3
[1] 18
```

Everything that's put between parentheses is carried out first.

You also can use basic operators on complex numbers. The complex() function, for example, allows you to construct a whole set of complex numbers based on a vector with real parts and a vector with imaginary parts. For more information, see the Help page for ?complex.

Using mathematical functions

In R, of course, you want to use more than just basic operators. R comes with a whole set of mathematical functions you'd find on a technical calculator as well, and then some more. Table 4-2 lists the ones that we think you'll use most often, but feel free to go on a voyage of discovery for others. All these functions are vectorized, so you can use them on complete vectors.

The possibilities of R go far beyond this small list of functions, however. We cover some of the special cases in the following sections.

Table 4-2	Useful Mathematical Functions in R
Function	*What It Does*
abs(x)	Takes the absolute value of *x*
log(x, base=y)	Takes the logarithm of *x* with base *y*; if base is not specified, returns the natural logarithm
exp(x)	Returns the exponential of *x*
sqrt(x)	Returns the square root of *x*
factorial(x)	Returns the factorial of *x* (*x*!)
choose(x, y)	Returns the number of possible combinations when drawing *y* elements at a time from *x* possibilities

Calculating logarithms and exponentials

In R, you can take the logarithm of the numbers from 1 to 3 like this:

```
> log(1:3)
[1] 0.0000000 0.6931472 1.0986123
```

Whenever you use one of these functions, R calculates the natural logarithm if you don't specify any base.

You calculate the logarithm of these numbers with base 6 like this:

```
> log(1:3, base = 6)
[1] 0.0000000 0.3868528 0.6131472
```

For the logarithms with bases 2 and 10, you can use the convenience functions log2() and log10().

You carry out the inverse operation of log() by using exp(). This last function raises e to the power mentioned between parentheses, like this:

```
> x <- log(1:3)
> exp(x)
```

Again, you can add a vector as an argument, because the log() function is also vectorized.

Putting the science in scientific notation

If you raise numbers to a power, the result will quickly become a very large number. For example, if you raise 1000 to the power of 2, you get a million. If you try that in R, you see the following:

```
> 1000^2
[1] 1e+06
```

This may look a bit weird if you aren't familiar with the scientific notation of numbers. *Scientific notation* allows you to represent a very large or very small number in a convenient way. The number is presented as a decimal and an exponent, separated by e. You get the number by multiplying the decimal by 10 to the power of the exponent. The number 13,300, for example, also can be written as 1.33×10^4, which is 1.33e4 in R:

```
> 1.33e4
[1] 13300
```

Likewise, 0.0412 can be written as 4.12×10^{-2}, which is `4.12e-2` in R:

```
> 4.12e-2
[1] 0.0412
```

R doesn't use scientific notation just to represent very large or very small numbers; it also understands scientific notation when you write it. You can use numbers written in scientific notation as though they were regular numbers, like so:

```
> 1.2e6 / 2e3
[1] 600
```

R automatically decides whether to print a number in scientific notation. Its decision to use scientific notation doesn't change the number itself, nor the accuracy of the calculation; it just saves some space.

Rounding numbers

Although R can calculate accurately to up to 16 digits, you don't always want to use that many digits. In this case, you can use a couple functions in R to round numbers. To round a number to two digits after the decimal point, for example, use the `round()` function as follows:

```
> round(123.456, digits = 2)
[1] 123.46
```

You also can use the `round()` function to round numbers to multiples of 10, 100, and so on. For that, you just add a negative number as the digits argument:

```
> round(-123.456, digits = -2)
[1] -100
```

If you want to specify the number of significant digits to be retained, regardless of the size of the number, you use the `signif()` function instead:

```
> signif(-123.456, digits = 4)
[1] -123.5
```

Both `round()` and `signif()` round numbers to the nearest possibility. So, if the first digit that's dropped is smaller than 5, the number is rounded down. If it's bigger than 5, the number is rounded up.

If the first digit that is dropped is exactly 5, R uses a rule that's common in programming languages: Always round to the nearest even number. `round(1.5)` and `round(2.5)` both return 2, for example, and `round(-4.5)` returns -4.

Contrary to round(), three other functions always round in the same direction:

- ✔ floor(x) rounds to the nearest integer that's smaller than x. So floor(123.45) becomes 123 and floor(-123.45) becomes –124.

- ✔ ceiling(x) rounds to the nearest integer that's larger than x. This means ceiling(123.45) becomes 124 and ceiling(-123.45) becomes –123.

- ✔ trunc(x) rounds to the nearest integer in the direction of 0. So trunc(123.65) becomes 123 and trunc(-123.65) becomes –123.

Using trigonometric functions

All trigonometric functions are available in R: the sine, cosine, and tangent functions and their inverse functions. You can find them on the Help page you reach by typing **?Trig**.

You may want to try to calculate the cosine of an angle of 120 degrees like this:

```
> cos(120)
[1] 0.814181
```

This code doesn't give you the correct result, however, because R always works with angles in radians, not in degrees. Pay attention to this fact; if you forget, the resulting bugs may bite you hard in the, er, leg.

Instead, use a special constant called pi. This constant has the value of — you guessed it — π (3.141592653589 . . .).

The correct way to calculate the cosine of an angle of 120 degrees, then, is this:

```
> cos(120 * pi / 180)
[1] -0.5
```

Calculating whole vectors

Sometimes the result of a calculation is dependent on multiple values in a vector. One example is the sum of a vector; when any value changes in the vector, the outcome is different. We call this complete set of functions and operators the *vector operations.* In "Powering Up Your Math," later in this chapter, you calculate cumulative sums and calculate differences between adjacent values in a vector with arithmetic vector functions. We discuss other vector operations in Chapter 7.

Actually, operators are also functions. But it's useful to draw a distinction between functions and operators, because operators are used differently from other functions. It helps to know, though, that operators can, in all cases, be treated just like any other function if you put the operator between backticks and add the arguments between parentheses, like this:

```
> '+'(2, 3)
[1] 5
```

This may be useful later on when you want to apply a function over rows, columns, or subsets of your data, as discussed in Chapters 9 and 13.

To infinity and beyond

In some cases, you don't have real values to calculate with. In most real-life data sets, in fact, at least a few values are missing. Also, some calculations have infinity as a result (such as dividing by zero) or can't be carried out at all (such as taking the logarithm of a negative value). Luckily, R can deal with all these situations.

Using infinity

To start exploring infinity in R, see what happens when you try to divide by zero:

```
> 2 / 0
[1] Inf
```

R correctly tells you the result is Inf, or infinity. Negative infinity is shown as -Inf. You can use Inf just as you use a real number in calculations:

```
> 4 - Inf
[1] -Inf
```

To check whether a value is finite, use the functions is.finite() and is.infinite(). The first function returns TRUE if the number is finite; the second one returns TRUE if the number is infinite. (We discuss the logical values TRUE and FALSE in the next section.)

R considers everything larger than the largest number a computer can hold to be infinity — on most machines, that's approximately 1.8×10^{308}. This definition of infinity can lead to unexpected results, as shown in the following example:

```
> is.finite(10^(305:310))
[1] TRUE TRUE TRUE TRUE FALSE FALSE
```

What does this line of code mean now? See whether you understand the nesting and vectorization in this example. If you break up the line starting from the inner parentheses, it becomes comprehensible:

✔ You know already that `305:310` gives you a vector, containing the integers from 305 to 310.

✔ All operators are vectorized, so `10^(305:310)` gives you a vector with the results of 10 to the power of 305, 306, 307, 308, 309, and 310.

✔ That vector is given as an argument to `is.finite()`. This function tells you that the two last results — `10^309` and `10^310` — are infinite for R.

Dealing with undefined outcomes

Your math teacher probably explained that if you divide any real number by infinity, you get zero. But what if you divide infinity by infinity?

```
> Inf / Inf
[1] NaN
```

Well, R tells you that the outcome is `NaN`. That result simply means *Not a Number*. This is R's way of telling you that the outcome of that calculation is not defined.

The funny thing is that R actually considers `NaN` to be numeric, so you can use `NaN` in calculations. The outcome of those calculations is always `NaN`, though, as you see here:

```
> NaN + 4
[1] NaN
```

You can test whether a calculation results in `NaN` by using the `is.nan()` function. Note that both `is.finite()` and `is.infinite()` return `FALSE` when you're testing on a `NaN` value.

Dealing with missing values

As we mention earlier in this chapter, one of the most common problems in statistics is incomplete data sets. To deal with missing values, R uses the reserved keyword `NA`, which stands for *Not Available*. You can use `NA` as a valid value, so you can assign it as a value as well:

```
> x <- NA
```

You have to take into account, however, that calculations with a value of NA also generally return NA as a result:

```
> x + 4
[1] NA
> log(x)
[1] NA
```

If you want to test whether a value is NA, you can use the is.na() function, as follows:

```
> is.na(x)
[1] TRUE
```

Note that the is.na() function also returns TRUE if the value is NaN. The functions is.finite(), is.infinite(), and is.nan() return FALSE for NA values.

Calculating infinite, undefined, and missing values

Table 4-3 provides an overview of results from the functions described in the preceding sections. You are unlikely to use any of these except for is.na(), which you may use quite a lot!

Table 4-3	Results of Infinite, NaN, and Missing Values			
Function	**Inf**	**–Inf**	**NaN**	**NA**
is.finite()	FALSE	FALSE	FALSE	FALSE
is.infinite()	TRUE	TRUE	FALSE	FALSE
is.nan()	FALSE	FALSE	TRUE	FALSE
is.na()	FALSE	FALSE	TRUE	TRUE

Organizing Data in Vectors

Vectors are the most powerful features of R, and in this section, you see why and how you use them.

A *vector* is a one-dimensional set of values, all the same type. It's the smallest unit you can work with in R. A single value is technically a vector as well — a vector with only one element. In mathematics vectors are almost always used with numerical values, but in R they also can include other types of data, like character strings (see Chapter 5).

Discovering the properties of vectors

Vectors have a structure and a type, and R is a bit sensitive about both. Feeding R the wrong type of vector is like trying to make your cat eat dog food: Something will happen, and chances are that it won't be what you hoped for. So, you'd better know what type of vector you have.

Looking at the structure of a vector

R gives you an easy way to look at the structure of any object. This method comes in handy whenever you doubt the form of the result of a function or a script you wrote. To take a peek inside R objects, use the `str()` function.

The `str()` function gives you the type and structure of the object.

Take a look at the vector `baskets.of.Granny`:

```
> str(baskets.of.Granny)
 num [1:6] 12 4 5 6 9 3
```

R tells you a few things here:

- First, it tells you that this is a `num` (numeric) type of vector.

- Next to the vector type, R gives you the dimensions of the vector. This example has only one dimension, and that dimension has indices ranging from 1 to 6.

- Finally, R gives you the first few values of the vector. In this example, the vector has only six values, so you see all of them.

If you want to know only how long a vector is, you can simply use the `length()` function, as follows:

```
> length(baskets.of.Granny)
 [1] 6
```

Vectors in R can have other types as well. If you look at the vector `authors`, for example (refer to Chapter 3), you see a small difference:

```
> authors <- c("Andrie", "Joris")
> str(authors)
 chr [1:2] "Andrie" "Joris"
```

Again, you get the dimensions, the range of the indices, and the values. But this time, R tells you the type of vector is `chr` (character).

In this book, we discuss the following types of vectors:

- ✔ **Numeric vectors,** containing all kinds of numbers.

- ✔ **Integer vectors,** containing integer values. (An integer vector is a special kind of numeric vector.)

- ✔ **Logical vectors,** containing logical values (TRUE and/or FALSE).

- ✔ **Character vectors,** containing text.

- ✔ **Datetime vectors,** containing dates and times in different formats.

- ✔ **Factors,** a special type of vector to work with categories.

All of the listed types of vectors may have missing values (NA).

We discuss the first three types in this chapter. You learn about character vectors and factors in Chapter 5, and about datetime vectors in Chapter 6.

R makes clear distinctions among these types of vectors, partly for reasons of logic. Multiplying two words, for example, doesn't make sense.

Testing vector types

Apart from the str() function, R contains a set of functions that allow you to test for the type of a vector. All these functions have the same syntax: is, a dot, and then the name of the type.

You can test whether a vector is of type foo by using the is.foo() function. This test works for every type of vector; just replace foo with the type you want to check.

To test whether baskets.of.Granny is a numeric vector, for example, use the following code:

```
> is.numeric(baskets.of.Granny)
[1] TRUE
```

You may think that baskets.of.Granny is a vector of integers, so check it, as follows:

```
> is.integer(baskets.of.Granny)
[1] FALSE
```

R disagrees with the math teacher here. *Integer* has a different meaning for R than it has for us. The result of is.integer() isn't about the value but about the way the value is stored in memory.

R has two main modes for storing numbers. The standard mode is double. In this mode, every number uses 64 bits of memory. The number also is stored in three parts. One bit indicates the sign of the number, 52 bits represent the decimal part of the number, and the remaining bits represent the exponent. This way, you can store numbers as big as 1.8×10^{308} in only 64 bits. The integer mode takes only 32 bits of memory, and the numbers are represented as binary integers in the memory. So, the largest integer is about 2.1 billion, or, more exactly, $2^{31} - 1$. That's 31 bits to represent the number itself, 1 bit to represent the sign of the number, and –1 because you start at 0.

You should use integers if you want to do exact integer calculations on small integers or if you want to save memory. Otherwise, the mode double works just fine. One of the nice things about R is that you hardly ever need to worry about whether something is stored as an integer or a double!

You force R to store a number as an integer by adding L after it, as in the following example:

```
> x <- c(4L, 6L)
> is.integer(x)
[1] TRUE
```

Whatever mode is used to store the value, is.numeric() returns TRUE in both cases.

Creating vectors

To create a vector from a simple sequence of integers, for example, you use the colon operator (:). The code 3:7 gives you a vector with the numbers 3 to 7, and 4:-3 creates a vector with the numbers 4 to –3, both in steps of 1. To make bigger or smaller steps in a sequence, use the seq() function. This function's by argument allows you to specify the amount by which the numbers should increase or decrease. For a vector with the numbers 4.5 to 2.5 in steps of -0.5, for example, you use the following code:

```
> seq(from = 4.5, to = 2.5, by = -0.5)
[1] 4.5 4.0 3.5 3.0 2.5
```

Alternatively, you can specify the length of the sequence by using the argument length.out. R calculates the step size itself. You make a vector of nine values going from –2.7 to 1.3 like this:

```
> seq(from = -2.7, to = 1.3, length.out = 9)
[1] -2.7 -2.2 -1.7 -1.2 -0.7 -0.2 0.3 0.8 1.3
```

You don't have to write out the argument names if you give the values for the arguments in the correct order. The code `seq(4.5, 2.5, -0.5)` does exactly the same things as `seq(from = 4.5, to = 2.5, by = -0.5)`. But if you use the argument `length.out`, you always have to spell it out.

Combining vectors

To dive a bit deeper into how you can use vectors, let's get back to our All-Star Grannies example (refer to "Using arithmetic operators," earlier in this chapter). You created two vectors that contain the number of baskets that Granny and her friend Geraldine scored in the six games of this basketball season:

```
> baskets.of.Granny <- c(12, 4, 4, 6, 9, 3)
> baskets.of.Geraldine <- c(5, 3, 2, 2, 12, 9)
```

The `c()` function stands for *combine*. It doesn't create vectors — it just combines them.

In the preceding examples, you give six values as arguments to the `c()` function and get one combined vector in return. As you know, R considers each value a vector with one element. You also can use the `c()` function to combine vectors with more than one value, as in the following example:

```
> all.baskets <-c(baskets.of.Granny, baskets.of.Geraldine)
> all.baskets
 [1] 12 4 4 6 9 3 5 3 2 2 12 9
```

The result of this code is a vector with all 12 values.

In this code, the `c()` function maintains the order of the numbers. This example illustrates a second important feature of vectors: Vectors have an order. This order turns out to be very useful when you need to manipulate the individual values in the vector, as you do in "Getting Values in and out of Vectors," later in this chapter.

Repeating vectors

You can combine a vector with itself if you want to repeat it, but if you want to repeat the values in a vector many times, using the `c()` function becomes a bit impractical. R makes life easier by offering you a function for repeating a vector: `rep()`.

You can use the rep() function in several ways. If you want to repeat the complete vector a set number of times, for example, you specify the argument times. To repeat the vector c(0, 0, 7) three times, use this code:

```
> rep(c(0, 0, 7), times = 3)
[1] 0 0 7 0 0 7 0 0 7
```

You also can repeat every value by specifying the argument each, like this:

```
> rep(c(2, 4, 2), each = 3)
[1] 2 2 2 4 4 4 2 2 2
```

R has a little trick up its sleeve. You can tell R for each value how often it has to be repeated. To take advantage of that magic, tell R how often to repeat each value in a vector by using the times argument:

```
> rep(c(0, 7), times = c(4, 2))
[1] 0 0 0 0 7 7
```

And you can, like in seq, use the argument length.out to tell R how long you want it to be. R will repeat the vector until it reaches that length, even if the last repetition is incomplete, like so:

```
> rep(1:3, length.out = 7)
[1] 1 2 3 1 2 3 1
```

Getting Values in and out of Vectors

Vectors would be pretty impractical if you couldn't look up and manipulate individual values. You can perform these tasks easily by using R's advanced, powerful indexing system.

Understanding indexing in R

Every time R shows you a vector, it displays a number such as [1] in front of the output. In this example, [1] tells you where the first position in your vector is. This number is called the *index* of that value. If you make a longer vector — say, with the numbers from 1 to 30 — you see more indices. Consider this example:

```
> numbers <- 30:1
> numbers
 [1] 30 29 28 27 26 25 24 23 22 21 20 19 18 17 16 15 14
[18] 13 12 11 10 9 8 7 6 5 4 3 2 1
```

Here, you see that R counts 13 as the 18th value in the vector. At the beginning of every line, R tells you the index of the first value in that line.

If you try this example on your computer, you may see a different index at the beginning of the line, depending on the width of your console.

Extracting values from a vector

Those brackets ([]) illustrate another strong point of R. Square brackets represent a function that you can use to extract a value from that vector. You can get the fifth value of the preceding number vector like this:

```
> numbers[5]
[1] 26
```

Okay, this example isn't too impressive, but the bracket function takes vectors as arguments. If you want to select more than one number, you can simply provide a vector of indices as an argument inside the brackets, like so:

```
> numbers[c(5, 11, 3)]
[1] 26 20 28
```

R returns a vector with the numbers in the order you asked for. So, you can use the indices to order the values the way you want.

You also can store the indices you want to retrieve in another vector and give that vector as an argument, as in the following example:

```
> indices <- c(5, 11, 3)
> numbers[indices]
[1] 26 20 28
```

You can use indices to drop values from a vector as well. If you want all the numbers except for the third value, you can do that with the following code:

```
> numbers[-3]
 [1] 30 29 27 26 25 24 23 22 21 20 19 18 17 16 15 14 13
[18] 12 11 10 9 8 7 6 5 4 3 2 1
```

Here, too, you can use a complete vector of indices. If you want to expel the first 20 numbers, use this code:

```
> numbers[-(1:20)]
 [1] 10 9 8 7 6 5 4 3 2 1
```

Be careful to add parentheses around the sequence. If you don't, R will interpret that as meaning the sequence from –1 to 20, which isn't what you want here. If you try that code, you get the following error message:

```
> numbers[-1:20]
Error in numbers[-1:20] : only 0's may be mixed with negative subscripts
```

This message makes you wonder what the index 0 is. Well, it's literally nothing. If it's the only value in the index vector, you get an empty, or zero-length, vector back, whatever sign you give it; otherwise, it won't have any effect.

You can't mix positive and negative index values, so either select a number of values or drop them.

You can do a lot more with indices — they help you write concise and fast code, as we show you in the following sections and chapters.

Changing values in a vector

Let's get back to the All-Star Grannies. In the previous sections, you created two vectors containing the number of baskets that Granny and Geraldine made in six basketball games.

But suppose that Granny tells you that you made a mistake: In the third game, she made five baskets, not four. You can easily correct this mistake by using indices, as follows:

```
> baskets.of.Granny[3] <- 5
> baskets.of.Granny
[1] 12 4 5 6 9 3
```

The assignment to a specific index is actually a function as well. It's different, however, from the brackets function (refer to "Extracting values from a vector," earlier in this chapter), because you also give the replacement values as an argument. Boring technical stuff, you say? Not if you realize that because the index assignment is a vectorized function, you can use recycling!

Imagine that you made two mistakes in the number of baskets that Granny's friend Geraldine scored: She actually scored four times in the second and fourth games. To correct the baskets for Geraldine, you can use the following code:

```
> baskets.of.Geraldine[c(2, 4)] <- 4
> baskets.of.Geraldine
[1] 5 4 2 4 12 9
```

How cool is that? You have to be careful, though. R doesn't tell you when it's recycling values, so a typo may give you unexpected results. Later in this chapter, you find out more about how recycling actually works.

R doesn't have an Undo button, so when you change a vector, there's no going back. You can prevent disasters by first making a copy of your object and then changing the values in the copy, as shown in the following example. First, make a copy by assigning the vector `baskets.of.Granny` to the object `Granny.copy`:

```
> Granny.copy <- baskets.of.Granny
```

You can check what's in both objects by typing the name on the command line and pressing Enter. Now you can change the vector `baskets.of.Granny`:

```
> baskets.of.Granny[4] <- 11
> baskets.of.Granny
[1] 12 4 5 11 9 3
```

If you make a mistake, simply assign the vector `Granny.copy` back to the object `baskets.of.Granny`, like this:

```
> baskets.of.Granny <- Granny.copy
> baskets.of.Granny
[1] 12 4 5 6 9 3
```

Working with Logical Vectors

Up to now, we haven't really discussed the values TRUE and FALSE. For some reason, the developers of R decided to call these values *logical values*. In other programming languages, TRUE and FALSE are known as *Boolean values*. As Shakespeare would ask, what's in a name? Whatever name they go by, these values come in handy when you start controlling the flow of your scripts, as we discuss in Chapter 9.

You can do a lot more with these values, however, because you can construct vectors that contain only logical values — the logical vectors that we mention in "Looking at the structure of a vector," earlier in this chapter. You can use these vectors as an argument for the index functions, which makes for a powerful tool.

Comparing values

To build logical vectors, you'd better know how to compare values, and R contains a set of operators that you can use for this purpose (see Table 4-4).

All these operators are, again, vectorized. You can compare two vectors in a single line of code. Or, since R recycles the arguments in this case, you can compare a complete vector with a single value in one line. In the continuing All-Star Grannies example, to find out which games Granny scored more than five baskets in, try:

```
> baskets.of.Granny > 5
[1] TRUE FALSE FALSE TRUE TRUE FALSE
```

You can see that the result is the first, fourth, and fifth games. This example works well for small vectors like this one, but if you have a very long vector, counting the number of games would be a hassle. For that purpose, R offers the delightful which() function. To find out which games Granny scored more than five baskets in, try:

```
> which(baskets.of.Granny > 5)
[1] 1 4 5
```

With this one line of code, you actually do two different things: First, you make a logical vector by checking every value in the vector to see whether it's greater than five. Then you pass that vector to the which() function, which returns the indices in which the value is TRUE.

Table 4-4	Comparing Values in R
Operator	*Result*
x == y	Returns TRUE if *x* exactly equals *y*
x != y	Returns TRUE if *x* differs from *y*
x > y	Returns TRUE if *x* is larger than *y*
x >= y	Returns TRUE if *x* is larger than or exactly equal to *y*
x < y	Returns TRUE if *x* is smaller than *y*
x <= y	Returns TRUE if *x* is smaller than or exactly equal to *y*
x & y	Returns the result of *x* and *y*
x | y	Returns the result of *x* or *y*
! x	Returns not *x*
xor(x, y)	Returns the result of *x* xor *y* (*x* or *y* but not *x* and *y*)

The which() function takes a logical vector as argument. Hence, you can save the outcome of a logical vector in an object and pass that to the which() function, as in the next example. You also can use all these operators to compare vectors value by value. You can easily find out the games in which Geraldine scored fewer baskets than Granny like this:

```
> the.best <- baskets.of.Geraldine < baskets.of.Granny
> which(the.best)
[1] 1 3 4
```

Always put spaces around the less than (<) and greater than (>) operators. Otherwise, R may mistake x < -3 for the assignment x <- 3. The difference may seem small, but it has a huge effect on the result. Technically, you also can use the equal sign (=) as an assignment to prevent this problem, but = also is used to assign values to arguments in functions. In general, <- is the preferred way to assign a value to an object (although some R programmers disagree). So, it's up to you. We use <- in this book.

Using logical vectors as indices

The index function doesn't take only numerical vectors as arguments; it also works with logical vectors. You can use these logical vectors very efficiently to select some values from a vector. If you use a logical vector to index, R returns a vector with only the values for which the logical vector is TRUE.

In the preceding section, a logical vector, the.best, tells you the games in which Granny scored more than Geraldine did. If you want to know how many baskets Granny scored in those games, you can use this code:

```
> baskets.of.Granny[the.best]
[1] 12 5 6
```

You might think this is overkill. After all, you can just use the logical vector to select these values. But there's a catch, as shown in the next example. If you want to keep only the values larger than 2 in a vector x, you could do that with the following code:

```
> x <- c(3, 6, 1, NA, 2)
> x[x > 2]
[1] 3 6 NA
```

Wait — what is that NA value doing there? Take a step back, and look at the result of x > 2:

```
> x > 2
[1] TRUE TRUE FALSE NA FALSE
```

If you have a missing value in your vector, any comparison returns NA for that value (refer to "Dealing with missing values," earlier in this chapter).

It may seem that this NA is translated into TRUE, but that isn't the case. If you give NA as a value for the index, R puts NA in that place as well. So, in this case, R keeps the first and second values of x, drops the third, adds one missing value, and drops the last value of x as well.

Combining logical statements

Life would be boring if you couldn't combine logical statements. If you want to test whether a number lies within a certain interval, for example, you want to check whether it's greater than the lowest value and less than the top value. Maybe you want to know the games in which Granny scored the fewest or the most baskets. For that purpose, R has a set of logical operators that — you guessed it — are nicely vectorized (refer to Table 4-4).

To illustrate, using the knowledge you have now, try to find out the games in which Granny scored the fewest baskets and the games in which she scored the most baskets:

1. **Create two logical vectors, as follows:**

```
> min.baskets <- baskets.of.Granny == min(baskets.of.Granny)
> max.baskets <- baskets.of.Granny == max(baskets.of.Granny)
```

min.baskets tells you whether the value is equal to the minimum, and max.baskets tells you whether the value is equal to the maximum.

2. **Combine both vectors with the OR operator (|), as follows:**

```
> min.baskets | max.baskets
[1] TRUE FALSE FALSE FALSE FALSE TRUE
```

This method actually isn't the most efficient way to find those values. You see how to do things like this more efficiently with the match() function in Chapter 13. But this example clearly shows you how vectorization works for logical operators.

The NOT operator (!) is another example of the great power of vectorization. The NA values in the vector x have caused some trouble already, so you might want to get rid of them. You know from "Dealing with undefined outcomes," earlier in this chapter, that you have to check whether a value is missing by using the is.na() function. But you need the values that are *not* missing values, so invert the logical vector by preceding it with the ! operator. To drop the missing values in the vector x, for example, use the following code:

```
> x[!is.na(x)]
[1] 3 6 2 1
```

Logical operators deal with NA values in the same way that arithmetic operators do: If an NA value is compared to anything, the result always is NA. It is a common mistake to try testing missing values with a command like this:

```
> x == NA
```

That won't work — you need to use is.na().

Summarizing logical vectors

You also can use logical values in arithmetic operations as well. In that case, R sees TRUE as 1 and FALSE as 0. This allows for some pretty interesting constructs.

Suppose that you're not really interested in finding out the games in which Granny scored more than Geraldine did, but you want to know how often that happened. You can use the numerical translation of a logical vector for that purpose in the sum() function, as follows:

```
> sum(the.best)
[1] 3
```

In addition, you have an easy way to figure out whether any value in a logical vector is TRUE. Very conveniently, the function that performs that task is called any(). To ask R whether Granny was better than Geraldine in any game, use this code:

```
> any(the.best)
[1] TRUE
```

This result is a bit unfair for Geraldine, so you may want to check whether Granny was better than Geraldine in *all* the games. The R function you use for this purpose is called — surprise, surprise — all(). To find out whether Granny was always better than Geraldine, use the following code:

```
> all(the.best)
[1] FALSE
```

Powering Up Your Math

As we suggest throughout this chapter, vectorization is the Holy Grail for every R programmer. Most beginners struggle a bit with that concept because vectorization isn't one little trick, but a way of coding. Using the indices and vectorized operators, however, can save you a lot of coding and calculation time — and then you can call a gang of powerful functions to quickly perform more complex mathematical operations.

Why are these functions so helpful? Maybe you're like us: We're lazy and impatient enough to try to translate our code into "something with vectors" as often as possible. We don't like to type too much, and we definitely don't like to wait for the results. If you can relate, read on.

Using arithmetic vector operations

A third set of arithmetic functions consists of functions in which the outcome is dependent on more than one value in the vector. Often, the idea behind these operations requires some form of looping over the different values in a vector. Summing all values in a vector with the sum() function is such an operation. You find an overview of the most important functions in Table 4-5.

Table 4-5	Vector Operations
Function	*What It Does*
sum(x)	Calculates the sum of all values in *x*
prod(x)	Calculates the product of all values in *x*
min(x)	Gives the minimum of all values in *x*
max(x)	Gives the maximum of all values in *x*
cumsum(x)	Gives the cumulative sum of all values in *x*
cumprod(x)	Gives the cumulative product of all values in *x*
cummin(x)	Gives the minimum for all values in *x* from the start of the vector until the position of that value
cummax(x)	Gives the maximum for all values in *x* from the start of the vector until the position of that value
diff(x)	Gives for every value the difference between that value and the next value in the vector

Summarizing a vector

You can tell quite a few things about a set of values with one number. If you want to know the minimum and maximum number of baskets Granny made, for example, you use the functions `min()` and `max()`:

```
> min(baskets.of.Granny)
[1] 3
> max(baskets.of.Granny)
[1] 12
```

To calculate the sum and the product of all values in the vector, use the functions `sum()` and `prod()`, respectively.

These functions also accept multiple vectors as input, separated by a comma. For example, to calculate the sum of all the baskets made by Granny and Geraldine, try:

```
> sum(baskets.of.Granny, baskets.of.Geraldine)
[1] 75
```

The same works for the other vector operations in this section.

As we discuss in "Dealing with missing values," earlier in this chapter, calculations with missing values always return `NA` as a result. The same is true for vector operations as well. R, however, gives you a way to simply discard the missing values by setting the argument `na.rm` to `TRUE`. Take a look at the following example:

```
> x <- c(3, 6, 2, NA, 1)
> sum(x)
[1] NA
> sum(x, na.rm = TRUE)
[1] 12
```

This argument works in `sum()`, `prod()`, `min()`, and `max()`.

If you have a vector that contains only missing values and you set the argument `na.rm` to `TRUE`, the outcome of these functions is set in such a way that it doesn't have any effect on further calculations. The sum of missing values is `0`, the product is `1`, the minimum is `Inf`, and the maximum is `-Inf`. R won't always generate a warning in such a case, though. Only in the case of `min()` and `max()` does R tell you that there were no non-missing arguments.

Cumulating operations

Suppose that after every game, you want to update the total number of baskets that Granny made during the season. After the second game, that's the total of the first two games; after the third game, it's the total of the first three games; and so on. In other words, you want to calculate the cumulative sum of the baskets Granny scored. You can make this calculation easily by using the function cumsum() as in the following example:

```
> cumsum(baskets.of.Granny)
[1] 12 16 21 27 36 39
```

In a similar way, cumprod() gives you the cumulative product. You also can get the cumulative minimum and maximum with the related functions cummin() and cummax(). To find the maximum number of baskets Geraldine scored up to any given game, you can use the following code:

```
> cummax(baskets.of.Geraldine)
[1] 5 5 5 5 12 12
```

These functions don't have an extra argument to remove missing values. Missing values are propagated through the vector, as shown in the following example:

```
> cummin(x)
[1] 3 3 2 NA NA
```

Calculating differences

The last function we'll discuss in this section calculates differences between adjacent values in a vector. You can calculate the difference in the number of baskets between every two games Granny played by using the following code:

```
> diff(baskets.of.Granny)
[1] -8 1 1 3 -6
```

You get five numbers back. The first one is the difference between the first and the second game, the second is the difference between the second and the third game, and so on.

The vector returned by diff() is always one element shorter than the original vector you gave as an argument.

The rule about missing values applies here, too. When your vector contains a missing value, the result from that calculation will be NA. So, if you calculate the difference with the vector x, you get the following result:

```
> diff(x)
[1] 3 -4 NA NA
```

Because the fourth element of x is NA, the difference between the third and fourth element and between the fourth and fifth element will be NA as well. Just like the cumulative functions, the diff() function doesn't have an argument to eliminate the missing values.

Recycling arguments

In Chapter 3 and earlier in this chapter, we mention recycling arguments. Take a look again at how you calculate the total amount of money Granny and Geraldine raised (see "Using arithmetic operators," earlier in this chapter) or how you combine the first names and last names of three siblings (see Chapter 3). Each time, you combine a vector with multiple values and one with a single value in a function. R applies the function, using that single value for every value in the vector. But recycling goes far beyond these examples.

Any time you give two vectors with unequal lengths to a recycling function, R repeats the shortest vector as often as necessary to carry out the task you asked it to perform. In the earlier examples, the shortest vector is only one value long.

Suppose you split up the number of baskets Granny made into two-pointers and three-pointers:

```
> Granny.pointers <- c(10, 2, 4, 0, 4, 1, 4, 2, 7, 2, 1, 2)
```

You arrange the numbers in such a way that for every game, first the number of two-pointers is given, followed by the number of three-pointers.

Now Granny wants to know how many points she's actually scored this season. You can calculate that easily with the help of recycling:

```
> points <- Granny.pointers * c(2, 3)
> points
 [1] 20 6 8 0 8 3 8 6 14 6 2 6
> sum(points)
[1] 87
```

Now, what did you do here?

1. **You made a vector with the number of points for each basket:**

   ```
   c(2, 3)
   ```

2. **You told R to multiply that vector by the vector `Granny.pointers`.**

   R multiplied the first number in Granny.pointers by 2, the second by 3, the third by 2 again, and so on.

3. You put the result in the variable `points`.

4. You summed all the numbers in `points` **to get the total number of points scored.**

In fact, you can just leave out Step 3. The nesting of functions allows you to do this in one line of code:

```
> sum(Granny.pointers * c(2, 3))
```

Recycling can be a bit tricky. If the length of the longer vector isn't exactly a multiple of the length of the shorter vector, you can get unexpected results.

Now Granny wants to know how much she improved every game. Being lazy, you have a cunning plan. With `diff()`, you calculate how many more or fewer baskets Granny made than she made in the game before. Then you use the vectorized division to divide these differences by the number of baskets in the game. To top it off, you multiply by 100 and round the whole vector. All these calculations take one line of code:

```
> round(diff(baskets.of.Granny) / baskets.of.Granny * 100 )
 1st 2nd 3rd 4th 5th 6th
 -67 25  20  50 -67 -267
```

That last value doesn't look right, because it's impossible to score more than 100 percent fewer baskets. R doesn't just give you that weird result; it also warns you that the length of `diff(baskets.of.Granny)` doesn't fit the length of `baskets.of.Granny`:

```
Warning message:
In diff(baskets.of.Granny) / baskets.of.Granny :
  longer object length is not a multiple of shorter object length
```

The vector `baskets.of.Granny` is six values long, but the outcome of `diff(baskets.of.Granny)` is only five values long. So the decrease of 267 percent is, in fact, the last value of `baskets.of.Granny` divided by the first value of `diff(baskets.of.Granny)`. In this example, the shortest vector, `diff(baskets.of.Granny)`, gets recycled by the division operator.

That result wasn't what you intended. To prevent that outcome, you should use only the first five values of `baskets.of.Granny`, so the length of both vectors match:

```
> round(diff(baskets.of.Granny) / baskets.of.Granny[1:5] * 100)
 2nd 3rd 4th 5th 6th
 -67  25  20  50 -67
```

And all that is vectorization.

Chapter 5

Getting Started with Reading and Writing

*I*t's not for no reason that reading and writing are considered to be two of the three Rs in elementary education (reading, 'riting, and 'rithmetic). In this chapter, you get to work with words in R.

You assign text to variables. You manipulate these variables in many different ways, including finding text within text and concatenating different pieces of text into a single vector. You also use R functions to sort text and to find words in text with some powerful pattern search functions, called regular expressions. Finally, you work with *factors,* the R way of representing categories (or categorical data, as statisticians call it).

Using Character Vectors for Text Data

Text in R is represented by character vectors. A *character vector* is — you guessed it! — a vector consisting of strings of characters. In Figure 5-1, you can see that each element of a character vector is a bit of text.

In the world of computer programming, text often is referred to as a *string.* In this chapter, we use the word *text* to refer to a single element of a vector, but you should be aware that the R Help files sometimes refer to *strings* and sometimes to *text.* These terms mean the same thing.

Figure 5-1:
Each
element of
a character
vector is a
bit of text,
also known
as a string.

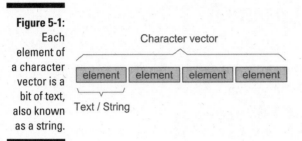

In this section, you take a look at how R uses character vectors to represent text. You assign some text to a character vector and get it to extract subsets of that data. You also get familiar with the very powerful concept of *named vectors,* vectors in which each element has a name. This is useful because you can then refer to the elements by name as well as position.

Assigning a value to a character vector

You assign a value to a character vector by using the assignment operator (<-), the same way you do for all other variables. You test whether a variable is of class character, for example, by using the is.character() function as follows:

```
> x <- "Hello world!"
> is.character(x)
TRUE
```

Notice that x is a character vector of length 1. To find out how many characters are in the text, use nchar():

```
> length(x)
[1] 1
> nchar(x)
[1] 12
```

The results tell you that x has length 1 and that the single element in x has 12 characters.

Creating a character vector with more than one element

To create a character vector with more than one element, use the combine function, c():

```
x <- c("Hello", "world!")
> length(x)
[1] 2
> nchar(x)
[1] 5 6
```

Notice that this time, R tells you that your vector has length 2 and that the first element has five characters and the second element has six characters.

Extracting a subset of a vector

You use the same indexing rules for character vectors that you use for numeric vectors (or for vectors of any type). The process of referring to a subset of a vector through indexing its elements is also called *subsetting*. In other words, subsetting is the process of extracting a subset of a vector.

To illustrate how to work with vectors, and specifically how to create subsets, we use the built-in datasets letters and LETTERS. Both are character vectors consisting of the letters of the alphabet, in lowercase (letters) and uppercase (LETTERS). Try it:

```
> letters
 [1] "a" "b" "c" "d" "e" "f" "g" "h" "i" "j" "k"
[12] "l" "m" "n" "o" "p" "q" "r" "s" "t" "u" "v"
[23] "w" "x" "y" "z"

> LETTERS
 [1] "A" "B" "C" "D" "E" "F" "G" "H" "I" "J" "K"
[12] "L" "M" "N" "O" "P" "Q" "R" "S" "T" "U" "V"
[23] "W" "X" "Y" "Z"
```

Aside from being useful to illustrate the use of subsets in this chapter, you can use these built-in vectors whenever you need to make lists of things.

Let's return to the topic of creating subsets. To extract a specific element from a vector, use square brackets. To get the tenth element of letters, for example, use the following:

```
> letters[10]
[1] "j"
```

To get the last three elements of LETTERS, use:

```
> LETTERS[24:26]
[1] "X" "Y" "Z"
```

The colon operator (:) in R is a handy way of creating sequences, so 24:26 results in 25, 25, 26. When this appears inside the square brackets, R returns elements 24 through 26.

In our last example, it was easy to extract the last three letters of LETTERS, because you know that the alphabet contains 26 letters. Quite often, you don't know the length of a vector. You can use the tail() function to display the trailing elements of a vector. To get the last five elements of LETTERS, try:

```
> tail(LETTERS, 5)
[1] "V" "W" "X" "Y" "Z"
```

Similarly, you can use the head() function to get the first element of a variable. By default, both head() and tail() returns six elements, but you can tell it to return any specific number of elements in the second argument. Try extracting the first ten letters:

```
> head(letters, 10)
[1] "a" "b" "c" "d" "e" "f" "g" "h" "i" "j"
```

Naming the values in your vectors

Until this point in the book, we've referred to the elements of vectors by their positions — that is, x[5] refers to the fifth element in vector x. One very powerful feature in R, however, gives names to the elements of a vector, which allows you to refer to the elements by name.

You can use these named vectors in R to associate text values (names) with any other type of value. Then you can refer to these values by name in addition to position in the list. This format has a wide range of applications — for example, named vectors make it easy to create lookup tables.

Looking at how named vectors work

To illustrate named vectors, take a look at the built-in dataset islands, a named vector that contains the surface area of the world's 48 largest land masses (continents and large islands). You can investigate its structure with str(), as follows:

```
> str(islands)
 Named num [1:48] 11506 5500 16988 2968 16...
 - attr(*, "names")= chr [1:48] "Africa" "Antarctica" "Asia" "Australia"...
```

R reports the structure of islands as a named vector with 48 elements. In the first line of the results of str(), you see the values of the first few elements of `islands`. On the second line, R reports that the named vector has an attribute containing `names` and reports that the first few elements of this attribute are `"Africa"`, `"Antarctica"`, `"Asia"`, and `"Australia"`.

Because each element in the vector has a value as well as a name, now you can subset the vector by name. To retrieve the sizes of Asia, Africa, and Antarctica, use the following:

```
> islands[c("Asia", "Africa", "Antarctica")]

      Asia     Africa Antarctica
     16988      11506       5500
```

You use the `names()` function to retrieve the names of a named vector:

```
> names(islands)[1:9]
[1] "Africa"     "Antarctica"  "Asia"
[4] "Australia"  "Axel Heiberg" "Baffin"
[7] "Banks"      "Borneo"      "Britain"
```

This function allows you to do all kinds of interesting things. Imagine you wanted to know the names of the six largest islands. To do this, you would retrieve the names of `islands` after sorting it in decreasing order:

```
> names(sort(islands, decreasing = TRUE)[1:6])
[1] "Asia"          "Africa"      "North America"
[4] "South America" "Antarctica"   "Europe"
```

Creating and assigning named vectors

You use the assignment operator (`<-`) to assign names to vectors in much the same way that you assign values to character vectors (see "Assigning a value to a character vector," earlier in this chapter).

Imagine you want to create a named vector with the number of days in each month. First, create a numeric vector containing the number of days in each month. Then use the built-in dataset `month.name` for the month names, as follows:

```
> month.days <- c(31, 28, 31, 30, 31, 30, 31, 31, 30, 31, 30, 31)
> names(month.days) <- month.name
> month.days
  January  February     March     April
       31        28        31        30
      May      June      July    August
       31        30        31        31
September   October  November  December
       30        31        30        31
```

Now you can use this vector to find the names of the months with 31 days:

```
> names(month.days[month.days == 31])

[1] "January"  "March"    "May"
[4] "July"     "August"   "October"
[7] "December"
```

This technique works because you subset month.days to return only those values for which month.days equals 31, and then you retrieve the names of the resulting vector.

The double equal sign (==) indicates a test for equality (see Chapter 4). Make sure not to use the single equal sign (=) for equality testing. Not only will a single equal sign not work, but it can have strange side effects because R interprets a single equal sign as an assignment. In other words, the operator = in many cases is the same as <-.

Manipulating Text

When you have text, you need to be able to manipulate it, for example by splitting or combining words. You also may want to analyze your text to find out whether it contains certain keywords or patterns.

In this section, you work with the string splitting and concatenation functions of R. *Concatenating* (combining) strings is something that programmers do very frequently. For example, when you create a report of your results, it's customary to combine descriptive text with the actual results of your analysis so that the reader of your results can easily digest it.

Finally, you start to work with finding words and patterns inside text, and you meet regular expressions, a powerful way of doing a wildcard search of text.

String theory: Combining and splitting strings

A collection of combined letters and words is called a *string*. Whenever you work with text, you need to be able to concatenate words (string them together) and split them apart. In R, you use the paste() function to concatenate and the strsplit() function to split. In this section, we show you how to use both functions.

Splitting text

First, create a character vector called `pangram`, and assign it the value `"The quick brown fox jumps over the lazy dog"`, as follows:

```
> pangram <- "The quick brown fox jumps over the lazy dog"
> pangram
[1] "The quick brown fox jumps over the lazy dog"
```

To split this text at the *word boundaries* (spaces), you can use `strsplit()` as follows:

```
> strsplit(pangram, " ")
[[1]]
[1] "The"    "quick" "brown" "fox"    "jumps" "over"  "the"   "lazy"  "dog"
```

Notice that the unusual first line of `strsplit()`'s output consists of `[[1]]`. Similar to the way that R displays vectors, `[[1]]` means that R is showing the first component of a *list*. Lists are extremely important concepts in R; they allow you to combine all kinds of variables. You can read more about lists in Chapter 7.

In the preceding example, this list has only a single component. Yes, that's right: The list has one component, but that component is a vector.

To extract a component from a list, you have to use double square brackets. Split your `pangram` into words, and assign the first component to a new variable called `words`, using double-square-brackets (`[[ ]]`) subsetting, as follows:

```
> words <- strsplit(pangram, " ")[[1]]
> words
[1] "The"    "quick" "brown" "fox"    "jumps" "over"  "the"   "lazy"  "dog"
```

To find the unique elements of a vector, including a vector of text, you use the `unique()` function. In the `words` object, `"the"` appears twice: once in lowercase and once with the first letter capitalized. To get a list of unique words, first convert `words` to lowercase and then use `unique()`:

```
> unique(tolower(words))
[1] "the"    "quick" "brown" "fox"    "jumps" "over"  "lazy"
[8] "dog"
```

Concatenating text

Now that you've split text, you can concatenate these components so that they again form a single text string.

Changing text case

To change some elements of `words` to uppercase, use the `toupper()` function:

```
> toupper(words[c(4, 9)])
[1] "FOX" "DOG"
```

To change text to lowercase, use `tolower()`:

```
> tolower("Some TEXT in Mixed CASE")
[1] "some text in mixed case"
```

To concatenate text, use the `paste()` function:

```
> paste("The", "quick", "brown", "fox")
[1] "The quick brown fox"
```

By default, `paste()` uses a blank space to concatenate the vectors. In other words, you separate components with spaces. This is because `paste()` takes an argument that specifies the separator. The default for the `sep` argument is a space (" ") — it defaults to separating components with a blank space, unless you tell it otherwise.

When you use `paste()`, or any function that accepts multiple arguments, make sure that you pass arguments in the correct format. Take a look at this example, but notice that this time there is a `c()` function in the code:

```
> paste(c("The", "quick", "brown", "fox"))
[1] "The"   "quick" "brown" "fox"
```

What's happening here? Why doesn't `paste()` paste the words together? The reason is that, by using `c()`, you passed a vector as a single argument to `paste()`. The `c()` function combines objects into a vector (or list). By default, `paste()` concatenates separate vectors — it doesn't collapse elements of a vector.

For the same reason, `paste(words)` results in the following:

```
[1] "The"   "quick" "brown" "FOX"   "jumps" "over"  "the"   "lazy"  "DOG"
```

The `paste()` function takes two optional arguments. The separator (`sep`) argument controls how different vectors get concatenated, and the `collapse` argument controls how a vector gets collapsed into itself, so to speak.

When you want to concatenate the elements of a vector by using `paste()`, you use the `collapse` argument, as follows:

```
> paste(words, collapse = " ")
[1] "The quick brown FOX jumps over the lazy DOG"
```

The `collapse` argument of `paste()` can take any character value. If you want to paste together text by using an underscore, use the following:

```
> paste(words, collapse = "_")
[1] "The_quick_brown_FOX_jumps_over_the_lazy_DOG"
```

The `paste()` function takes vectors as input and joins them together. If one vector is shorter than the other, R *recycles* (repeats) the shorter vector to match the length of the longer one — a powerful feature.

Suppose that you have five objects, and you want to label them `"Sample 1"`, `"Sample 2"`, and so on. You can do this by passing a short vector with the value `Sample` and a long vector with the values `1:5` to `paste()`. In this example, the shorter vector is repeated five times:

```
> paste("Sample", 1:5)
[1] "Sample 1" "Sample 2" "Sample 3" "Sample 4" "Sample 5"
```

You can use `sep` and `collapse` in the same `paste` call. In this case, the vectors are first pasted with `sep` and then collapsed with `collapse`. Try this:

```
> paste(LETTERS[1:5], 1:5, sep = "_", collapse = "---")
[1] "A_1---B_2---C_3---D_4---E_5"
```

What happens here is that you first concatenate the elements of each vector with an underscore (that is, `A_1`, `B_2`, and so on), and then you collapse the results into a single string with `---` between each element.

Recycling character vectors

When you perform operations on vectors of different lengths, R automatically adjusts the length of the shorter vector to match the longer one. This is called *recycling,* since R *recycles* the element of the shorter vector to create a new vector that matches the original long vector.

This feature is very powerful but can lead to confusion if you aren't aware of it.

The rules for recycling character vectors are exactly the same as for numeric vectors (see Chapter 4).

Here are a few examples of vector recycling using `paste`:

```
> paste(c("A", "B"), c(1, 2, 3, 4),
          sep = "-")
[1] "A-1" "B-2" "A-3" "B-4"

> paste(c("A"), c(1, 2, 3, 4, 5),
          sep = "-")
[1] "A-1" "A-2" "A-3" "A-4" "A-5"
```

See how in the first example A and B get recycled to match the vector of length four. In the second example, the single A also gets recycled — in this case, five times.

Sorting text

What do league tables, telephone directories, dictionaries, and the index pages of a book have in common? They present data in some ordered manner. Data can be sorted alphabetically or numerically, in ascending or descending order. Like any programming language, R makes it easy to compile lists of sorted and ordered data.

Because text in R is represented as character vectors, you can sort these vectors using the same functions as you use with numeric data. For example, to get R to sort the alphabet in reverse, use the `sort()` function:

```
> sort(letters, decreasing = TRUE)
 [1] "z" "y" "x" "w" "v" "u" "t" "s" "r" "q" "p"
[12] "o" "n" "m" "l" "k" "j" "i" "h" "g" "f" "e"
[23] "d" "c" "b" "a"
```

Here you used the `decreasing` argument of `sort()`.

The `sort()` function sorts a vector. It doesn't sort the characters of each element of the vector. In other words, `sort()` doesn't mangle the word itself. You can still read each of the words in `words`.

Try it on your vector `words` that you created in the previous paragraph:

```
> sort(words)
[1] "brown" "DOG"   "FOX"   "jumps" "lazy"
[6] "over"  "quick" "the"   "The"
```

R performs *lexicographic* sorting, as opposed to, for example, the C language, which sorts in ASCII order. This means that the sort order will depend on the locale of the machine the code runs on. In other words, the sort order may be different if the machine running R is configured to use Danish than it will if the machine is configured to use English. The R help file contains this description:

> Beware of making any assumptions about the collation order: e.g., in Estonian, Z comes between S and T, and collation is not necessarily character-by-character — in Danish aa sorts as a single letter, after z.

In most cases, lexicographic sorting simply means that the sort order is independent of whether the string is in lowercase or uppercase. For more details, read the help text in `?sort` as well as `?Comparison`.

You can get help on any function by typing a question mark followed by the function name into the console. For other ways of getting help, refer to Chapter 11.

Finding text inside text

When you're working with text, often you can solve problems if you're able to find words or patterns inside text. Imagine you have a list of the states in the United States, and you want to find out which of these states contains the word *New*. Or, say you want to find out which state names consist of two words.

To solve the first problem, you need to search for individual words (in this case, the word *New*). And to solve the second problem, you need to search for multiple words. In this section, you solve both types of problem.

Searching for individual words

To investigate this problem, use the built-in dataset `state.name`, which contains — you guessed it — the names of the states of the United States:

```
> head(state.name)
[1] "Alabama"   "Alaska"     "Arizona"
[4] "Arkansas"  "California" "Colorado"
```

Broadly speaking, you can find substrings in text in two ways:

- ✔ **By position:** For example, you can tell R to get three letters starting at position 5.

- ✔ **By pattern:** For example, you can tell R to get substrings that match a specific word or pattern.

  A pattern works a bit like a wildcard. In some card games, you may use the Joker card to represent any other card. Similarly, a pattern in R can contain words or certain symbols with special meanings.

Searching by position

If you know the exact position of a subtext inside a text string, you use the `substr()` function to return the value. To extract the subtext that starts at the third position and stops at the sixth position of `state.name`, use:

```
> head(substr(state.name, start = 3, stop = 6))
[1] "abam" "aska" "izon" "kans" "lifo" "lora"
```

Searching by pattern

To find substrings, you can use the `grep()` function, which takes two essential arguments:

- ✔ `pattern`: The pattern you want to find.
- ✔ `x`: The character vector you want to search.

Suppose you want to find all the states that contain the pattern New. Do it like this:

```
> grep("New", state.name)
[1] 29 30 31 32
```

The result of grep() is a numeric vector with the positions of each of the components that contain the matching pattern. In other words, the 29th component of state.name contains the word *New*.

```
> state.name[29]
New Hampshire
```

Phew, that worked! But typing in the position of each matching text is going to be a lot of work. Fortunately, you can use the results of grep() directly to subset the original vector. You can do this by adding the argument value = TRUE. Try this:

```
> grep("New", state.name, value = TRUE)
[1] "New Hampshire" "New Jersey"
[3] "New Mexico"    "New York"
```

The grep() function is case sensitive — it only matches text in the same case (uppercase or lowercase) as your search pattern. If you search for the pattern "new" in lowercase, your search results are empty:

```
> grep("new", state.name, value = TRUE)
character(0)
```

Getting a grip on grep

The name of the grep() function originated in the Unix world. It's an acronym for Global Regular Expression Print. Regular expressions are a very powerful way of expressing patterns of matching text, usually in a very formal language. Whole books have been written about regular expressions. We give a very short introduction in "Revving up with regular expressions," later in this chapter.

The function name grep() appears in many programming languages that deal with text and reporting. Perl, for example, is famous for its extensive grep functionality. For more information, check out *Perl For Dummies*, 4th Edition, by Paul Hoffman (Wiley).

Searching for multiple words

So, how do you find the names of all the states with more than one word? This is easy when you realize that you can frame the question by finding all those states that contain a space:

```
> state.name[grep(" ", state.name)]
 [1] "New Hampshire"   "New Jersey"
 [3] "New Mexico"      "New York"
 [5] "North Carolina"  "North Dakota"
 [7] "Rhode Island"    "South Carolina"
 [9] "South Dakota"    "West Virginia"
```

The results include all the states that have two-word names, such as New Jersey, New York, North Carolina, South Dakota, and West Virginia.

You can see from this list that there are no state names that contain *East.* You can confirm this by doing another find:

```
> state.name[grep("East", state.name)]
character(0)
```

When the result of a character operation is an empty vector (that is, there is nothing in it), R represents it as `character(0)`. Similarly, an empty, or zero-length, numeric vector is represented with `integer(0)` or `numeric(0)` (see Chapter 4).

R makes a distinction between `NULL` and an empty vector. `NULL` usually means something is undefined. This is subtly different from something that is empty. For example, a character vector that happens to have no elements is still a character vector, represented by `character(0)`.

Substituting text

The `sub()` function (short for *substitute*) searches for a pattern in text and replaces this pattern with replacement text. You use `sub()` to substitute text for text, and you use its cousin `gsub()` to substitute all occurrences of a pattern. (The `g` in `gsub()` stands for *global.*)

Suppose you have the sentence *A wolf in cheap clothing,* which is clearly a mistake. You can fix it with a `gsub()` substitution. The `gsub()` function takes three arguments: the pattern to find, the replacement pattern, and the text to modify:

```
> gsub("cheap", "sheep's", "A wolf in cheap clothing")
[1] "A wolf in sheep's clothing"
```

Another common type of problem that can be solved with text substitution is removing substrings. Removing substrings is the same as replacing the substring with empty text (that is, nothing at all).

Imagine a situation in which you have three file names in a vector: `file_a.csv`, `file_b.csv`, and `file_c.csv`. Your task is to extract a, b, and c from those file names. You can do this in two steps: First, replace the pattern `"file_"` with nothing, and then replace the `".csv"` with nothing. You get your desired vector:

```
> x <- c("file_a.csv", "file_b.csv", "file_c.csv")
> y <- gsub("file_", "", x)
> y
[1] "a.csv" "b.csv" "c.csv"
> gsub("\\.csv", "", y)
[1] "a" "b" "c"
```

A dot (.) is a wildcard in a regular expression. It indicates "any character." If you want to refer to a point, you have to escape it with two backslashes. You can find more information on special characters in the section "Revving up with regular expressions" later in this chapter, as well as the Help for `?regex`.

Revving up with regular expressions

Until this point, you've worked mostly with fixed expressions to find or substitute text. This is useful but also limited. R supports the concept of regular expressions, which allows you to search for patterns inside text. (Strictly speaking, you already encountered `\\.` — also a regular expression.)

You may never have heard of regular expressions, but you're probably familiar with similar concepts. For example, if you've ever used an * or a ? to indicate any letter in a word, then you've used a form of wildcard search. Regular expressions support the idea of wildcards and much more.

Regular expressions allow three ways of making a search pattern more general than a single, fixed expression:

✔ **Alternatives:** You can search for instances of one pattern or another, indicated by the | symbol. For example `beach|beech` matches both *beach* and *beech*.

On English and American English keyboards, you can usually find the | on the same key as backslash (\).

✔ **Grouping:** You group patterns together using parentheses ( ). For example you write be(a|e)ch to find both *beach* and *beech*.

✔ **Quantifiers:** You specify whether a component in the pattern must be repeated or not by adding * (occurs zero or many times) or + (occurs one or many times). For example, to find either *bach* or *beech* (zero or more of *a* and *e* but not both), you use b(e*|a*)ch.

Extending text functionality with stringr

After this quick tour through the text manipulation functions of R, you probably wonder why all these functions have such unmemorable names and seemingly diverse syntax. If so, you're not alone. In fact, Hadley Wickham wrote a package available from CRAN that simplifies and standardizes working with text in R. This package is called stringr, and you can install it by using the R console or by choosing Tools ⇨ Install Packages. . . in RStudio (see Chapter 3).

Remember: Although you have to install a package only once, you have to load it into the workspace using the library() function every time you start a new R session and plan to use the functions in that package.

```
> install.packages("stringr")
> library("stringr")
```

Here are some of the advantages of using stringr rather than the standard R functions:

✔ **Function names and arguments are consistent and more descriptive.** For example, all stringr functions have names starting with str_ (such as str_detect() and str_replace()).

✔ stringr **has a more consistent way of dealing with cases with missing data or empty values.**

✔ stringr **has a more consistent way of ensuring that input and output data are of the same type.**

The stringr equivalent for grep() is str_detect(), and the equivalent for gsub() is str_replace_all().

As a starting point to explore stringr, you may find some of these functions useful:

✔ str_detect(): Detects the presence or absence of a pattern in a string

✔ str_extract(): Extracts the first piece of a string that matches a pattern

✔ str_length(): Returns the length of a string (in characters)

✔ str_locate(): Locates the position of the first occurrence of a pattern in a string

✔ str_match(): Extracts the first matched group from a string

✔ str_replace(): Replaces the first occurrence of a matched pattern in a string

✔ str_split(): Splits up a string into a variable number of pieces

✔ str_sub(): Extracts substrings from a character vector

✔ str_trim(): Trims white space from the start and end of string

✔ str_wrap(): Wraps strings into nicely formatted paragraphs

Try the following examples. First, create a new variable with five words:

```
> rwords <- c("bach", "back", "beech", "beach", "black")
```

Find either *beach* or *beech* using alternative matching:

```
> grep("beach|beech", rwords)
[1] 3 4
```

This means the search string was found in components 3 and 4 of rwords. To extract the actual components, you can use subsetting with square brackets:

```
> rwords[grep("beach|beech", rwords)]
[1] "beech" "beach"
```

Now use the grouping rule to extract the same words:

```
> rwords[grep("be(a|e)ch", rwords)]
[1] "beech" "beach"
```

Lastly, use the quantifier modification to extract *bach* and *beech* but not *beach:*

```
rwords[grep("b(e*|a*)ch", rwords)]
[1] "bach"  "beech"
```

To find more help in R about regular expressions, look at the Help page ?regex. Some other great resources for learning more about regular expressions are Wikipedia (http://en.wikipedia.org/wiki/Regular_expression) and www.regular-expressions.info, where you can find a quick-start guide and tutorials.

Factoring in Factors

In real-world problems, you often encounter data that can be described using words rather than numerical values. For example, cars can be red, green, or blue (or any other color); people can be left-handed or right-handed, male or female; energy can be derived from coal, nuclear, wind, or wave power. You can use the term *categorical data* to describe these examples — or anything else that can be classified in categories.

R has a special data structure for categorical data, called *factors*. Factors are closely related to characters because any character vector can be represented by a factor.

Factors are special types of objects in R. They're neither character vectors nor numeric vectors, although they have some attributes of both. Factors behave a little bit like character vectors in the sense that the unique categories are often text. Factors also behave a little bit like integer vectors because R encodes the levels as integers.

Creating a factor

To create a factor in R, you use the factor() function. The first three arguments of factor() warrant some exploration:

- ✔ x: The input vector that you want to turn into a factor.
- ✔ levels: An optional vector of the values that x might have taken. The default is lexicographically sorted, unique values of x.
- ✔ labels: Another optional vector that, by default, takes the same values as levels. You can use this argument to rename your levels, as we explain in the next paragraph.

The fact that you can supply both levels and labels to factor() can lead to confusion. Just remember that levels refers to the input values of x, while labels refers to the output values of the new factor.

Consider the following example of a vector consisting of compass directions:

```
> directions <- c("North", "East", "South", "South")
```

Notice that this vector contains the value "South" twice and lacks the value "West". First, convert directions to a factor:

```
> factor(directions)
[1] North East  South South
Levels: East North South
```

Notice that the levels of your new factor does not contain the value "West", which is as expected. In practice, however, it makes sense to have all the possible compass directions as levels of your factor. To add the missing level, you specify the levels arguments of factor():

```
> factor(directions, levels= c("North", "East", "South", "West"))
[1] North East  South South
Levels: North East South West
```

As you can see, the values are still the same but this time the levels also contain "West".

Now imagine that you actually prefer to have abbreviated names for the levels. To do this, you make use of the `labels` argument:

```
> factor(directions,
        levels = c("North", "East", "South", "West"),
        labels = c("N", "E", "S", "W"))
[1] N E S S
Levels: N E S W
```

Converting a factor

Sometimes you need to explicitly convert factors to either text or numbers. To do this, you use the functions `as.character()` or `as.numeric()`.

First, convert your `directions` vector into a factor called `directions.factor` (as you saw earlier):

```
> directions <- factor(c("North", "East", "South", "South"))
> directions
[1] North East  South South
Levels: East North South
```

Use `as.character()` to convert a factor to a character vector:

```
> as.character(directions)
[1] "North" "East"  "South" "South"
```

Use `as.numeric()` to convert a factor to a numeric vector. Note that this will return the numeric codes that correspond to the factor levels. For example, `"East"` corresponds to 1, `"North"` corresponds to 2, and so forth:

```
> as.numeric(directions)
[1] 2 1 3 3
```

Be very careful when you convert factors with numeric levels to a numeric vector. The results may not be what you expect.

For example, imagine you have a vector that indicates some test score results with the values `c(9, 8, 10, 8, 9)`, which you convert to a factor:

```
> numbers <- factor(c(9, 8, 10, 8, 9))
```

To look at the internal representation of `numbers`, use `str()`:

```
> str(numbers)
 Factor w/ 3 levels "8","9","10": 2 1 3 1 2
```

This indicates that R stores the values as `c(2, 1, 3, 1, 2)` with associated levels of `c("8", "9", "10")`. Figure 5-2 gives a graphical representation of this difference between the levels and the internal representation.

Figure 5-2:
A visual comparison between a numeric vector and a factor.

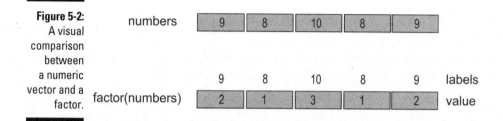

If you want to convert `numbers` to a character vector, the results are pretty much as you would expect:

```
> as.character(numbers)
[1] "9"  "8"  "10" "8"  "9"
```

However, if you simply use `as.numeric()`, your result is a vector of the internal level representations of your factor and not the original values:

```
> as.numeric(numbers)
[1] 2 1 3 1 2
```

The Help at `?factor` describes a solution to this problem. The solution is to first transform the factor to a character vector, and then to a numeric vector, like this:

```
> as.numeric(as.character(numbers))
[1]  9  8 10  8  9
```

This is an example of *nested functions* in R, in which you pass the results of one function to a second function. Nested functions are a bit like the Russian nesting dolls, where each toy is inside the next:

✔ The inner function, `as.character(numbers)`, contains the text `c("8", "9", "10")`.

✔ The outer function, `as.numeric(...)`, does the final conversion to `c(9, 8, 10, 8, 9)`.

Looking at levels

To look a little bit under the hood of the structure of a factor, use the `str()` function:

```
> str(state.region)
 Factor w/ 4 levels "Northeast","South",..: 2 4 4 2 4 4 1 2 2 2...
```

R reports the structure of `state.region` as a factor with four levels. You can see that the first two levels are `"Northeast"` and `"South"`, but these levels are represented as integers 1, 2, 3, and 4.

Factors are a convenient way to describe categorical data. Internally a factor is stored as a numeric value associated with each level. This means you can set and investigate the levels of a factor separately from the values of the factor.

To look at the levels of a factor, you use the `levels()` function. For example, to extract the factor levels of `state.region`, use the following:

```
> levels(state.region)
[1] "Northeast"     "South"        "North Central" "West"
```

Because the values of the factor are linked to the levels, when you change the levels, you also indirectly change the values themselves. To make this clear, change the levels of `state.region` to the values `"NE"`, `"S"`, `"NC"`, and `"W"`:

```
> levels(state.region) <- c("NE", "S", "NC", "W")
> head(state.region)
[1] S W W S W W
Levels: NE S NC W
```

Sometimes it's useful to know the number of levels of a factor. The convenience function `nlevels()` extracts the number of levels from a factor:

```
> nlevels(state.region)
[1] 4
```

Because the levels of a factor are internally stored by R as a vector, you also can extract the number of levels using `length`:

```
> length(levels(state.region))
[1] 4
```

For the very same reason, you can index the levels of a factor using standard vector subsetting rules. For example, to extract the second and third factor levels, use the following:

```
> levels(state.region)[2:3]
[1] "S"  "NC"
```

Distinguishing data types

In the field of statistics, being able to distinguish between variables of different types is very important. The type of data very often determines the type of analysis that can be performed. As a result, R offers the ability to explicitly classify data as follows:

- **Nominal data:** This type of data, which you represent in R using factors, distinguishes between different categories, but there is no implied order between categories. Examples of nominal data are colors (red, green, blue), gender (male, female), and nationality (British, French, Japanese).

- **Ordinal data:** Ordinal data is distinguished by the fact that there is some kind of natural order between elements but no indication of the relative size difference. Any kind of data that is possible to rank in order but not give exact values to is ordinal. For example, *low < medium < high* describes data that is ordered with three levels.

    In market research, it's very common to use a five-point scale to measure perceptions: *strongly disagree < disagree < neutral < agree < strongly agree*. This is also an example of ordinal data.

    Another example is the use of the names of colors to indicate order, such as *red < amber < green* to indicate project status.

    In R, you use *ordered factors* to describe ordinal data. For more on ordered factors, see the "Working with ordered factors" section, later in this chapter.

- **Numeric data:** You have numeric data when you can describe your data with numbers (for example, length, weight, or count). Numeric data has two subcategories.

    - **Interval scaled data:** You have interval scaled data when the interval between adjacent units of measurement is the same, but the zero point is arbitrary. An everyday example of interval scaled data is our calendar system. Each year has the same length, but the zero point is arbitrary. In other words, time didn't start in the year zero — we simply use a convenient year to start counting. This means you can add and subtract dates (and all other types of interval scaled data), but you can't meaningfully divide dates. Other examples include temperature, as well as anything else where there can be disagreement about where the starting point is.

Other examples of interval scaled data can be found in social science research such as market research.

In R you can use *integer* or *numeric* objects to represent interval scaled data.

- **Ratio scaled data:** This is data where all kinds of mathematical operations are allowed, in particular the ability to multiply and divide (in other words, take ratios). Most data in physical sciences are ratio scaled — for example, length, mass, and speed. In R, you use *numeric* objects to represent ratio scaled data.

Working with ordered factors

Sometimes data has some kind of natural order in which some elements are in some sense "better" or "worse" than other elements, but at the same time it's impossible to ascribe a meaningful value to these. An example is any situation where project status is described as low, medium, or high. A similar example is a traffic light that can be red, yellow, or green.

The name for this type of data, where rank ordering is important, is *ordinal data*. In R, there is a special data type for ordinal data. This type is called *ordered factors* and is an extension of factors that you're already familiar with.

To create an ordered factor in R, you have two options:

- Use the `factor()` function with the argument `ordered=TRUE`.
- Use the `ordered()` function.

Summarizing categorical data

In most practical cases where you have categorized data, some values are repeated. As a practical example, consider the states of the United States. Each state is in one of four regions: Northeast, South, North Central, or West (at least according to R). Have a look at the built-in dataset `state.region`:

```
> head(state.region)
[1] South West  West  South West  West
Levels: Northeast South North Central West
```

You can use the handy `table()` function to get a tabular summary of the values of a factor:

```
> table(state.region)
state.region
   Northeast    South North Central    West
           9       16            12      13
```

This tells you that the Northeast region has 9 states, the South region has 16 states, and so on.

The `table()` function works by counting the number of occurrences of each factor level. You can learn more about `table()` in the Help page at `?table`.

Say you want to represent the status of five projects. Each project has a status of low, medium, or high:

```
> status <- c("Lo", "Hi", "Med", "Med", "Hi")
```

Now create an ordered factor with this status data:

```
> ordered.status <- factor(status,
+                          levels = c("Lo", "Med", "Hi"),
+                          ordered = TRUE)
> ordered.status
[1] Lo  Hi  Med Med Hi
Levels: Lo < Med < Hi
```

REMEMBER

You can tell an ordered factor from an ordinary factor by the presence of directional signs (< or >) in the levels.

TIP

In R, there is a really big practical advantage to using ordered factors. A great many R functions recognize and treat ordered factors differently by printing results in the order that you expect. For example, compare the results of table(status) with table(ordered.status):

```
> table(status)
status
 Hi  Lo Med
  2   1   2
```

Notice that the results are ordered alphabetically. However, the results of performing the same function on the ordered factor yields results that are easier to interpret because they're now sorted in the order Lo, Med, Hi:

```
> table(ordered.status)
ordered.status
 Lo Med  Hi
  1   2   2
```

R preserves the ordering information inherent in ordered factors. In Part V, you see how this becomes an essential tool to gain control over the appearance of bar charts.

Also, in statistical modeling, R applies the appropriate statistical transformation (or contrasts) when you have factors or ordered factors in your model. In Chapter 15, you do some statistical modeling with categorical variables.

Chapter 6

Going on a Date with R

. . . .

. . . .

*A*ll kinds of real-world data are associated with a specific date or instant in time. Companies report results each quarter. Stock markets report closing prices daily. Network analysts measure traffic by the hour (if not by the minute). And of course, scientists measure air temperature, sometimes by the minute, sometimes by the day, and have done so for decades.

Dealing with dates accurately can be a complicated task. You have to account for time-zone differences, leap years, and regional differences in holidays. In addition, people report data differently in different places. For example, what an American would write as "May 12, 2010" or "05-12-10" would be written by someone from the United Kingdom as "12 May 2010" or "12-05-10." Working with a time instant on a specific day isn't any easier. The same time may be written as 9:25 p.m., 21:25, or 21h25 — not to mention time zones!

In this chapter, you look at the different ways of representing dates and times using R. You take control of the format of dates and time for pretty printing. Then you do some math with dates — addition and subtraction. Finally, you use some tricks to extract specific components, such as the month, from a date.

Working with Dates

R has a range of functions that allow you to work with dates and times. The easiest way of creating a date is to use the as.Date() function. For example, you write the opening day of the 2016 Rio Olympic Games as:

```
> xd <- as.Date("2016-08-05")
> xd
[1] "2016-08-05"
> str(xd)
 Date[1:1], format: "2016-08-05"
```

This works because the default format for dates in as.Date() is YYYY-MM-DD — four digits for year, and two digits for month and day, separated by a hyphen. In the next section, you get to specify dates in different formats.

To find out what day of the week this is, use weekdays():

```
> weekdays(xd)
[1] "Friday"
```

You can add or subtract numbers from dates to create new dates. For example, to calculate the date that is seven days in the future, use the following:

```
> xd + 7
[1] "2016-08-12"
```

In the same way as with numbers or text, you can put multiple dates into a vector. To create a vector of seven days starting on July 27, add 0:6 to the starting date. (*Remember:* The colon operator generates integer sequences.)

```
> xd + 0:6
[1] "2016-08-05" "2016-08-06" "2016-08-07" "2016-08-08"
[5] "2016-08-09" "2016-08-10" "2016-08-11"
```

Because the weekdays() function takes vector input, it returns the days of the week for this sequence:

```
> weekdays(xd + 0:6)
[1] "Friday"    "Saturday"  "Sunday"    "Monday"
[5] "Tuesday"   "Wednesday" "Thursday"
```

You can use the seq() function to create sequences of dates in a far more flexible way. As with numeric vectors, you have to specify at least three of the arguments (from, to, by, and length.out). However, in the case of Date objects, the by argument is very flexible. You specify by as a string consisting of a number followed by days, weeks, or months. Imagine you want to create a sequence of every second month of 2016, starting at January 1:

```
> startDate <- as.Date("2016-01-01")
> xm <- seq(startDate, by = "2 months", length.out = 6)
> xm
[1] "2016-01-01" "2016-03-01" "2016-05-01" "2016-07-01"
[5] "2016-09-01" "2016-11-01"
```

In addition to weekdays(), you also can get R to report on months() and quarters():

```
> months(xm)
[1] "January"   "March"     "May"       "July"
[5] "September" "November"
```

```
> quarters(xm)
[1] "Q1" "Q1" "Q2" "Q3" "Q3" "Q4"
```

The results of many date functions, including weekdays() and months() depends on the locale of the machine you're working on. The locale describes elements of international customization on a specific installation of R. This includes date formats, language settings, and currency settings. To find out some of the locale settings on your machine, use Sys.localeconv(). R sets the value of these variables at install time by interrogating the operating system for details. You can change these settings at runtime or during the session with Sys.setlocale().

To view the locale settings on your machine, try the following:

```
> Sys.localeconv()
```

Table 6-1 summarizes some useful functions for working with dates.

Table 6-1	**Useful Functions with Dates**
Function	*Description*
as.Date()	Converts character string to Date
weekdays()	Full weekday name in the current locale (for example, Sunday, Monday, Tuesday)
months()	Full month name in the current locale (for example, January, February, March)
quarters()	Quarter numbers (Q1, Q2, Q3, or Q4)
seq()	Generates dates sequences if you pass it a Date object as its first argument

Presenting Dates in Different Formats

You've probably noticed that as.Date() is fairly prescriptive in its defaults: It expects the date to be formatted in the order of year, month, and day. Fortunately, R allows you flexibility in specifying the date format.

By using the format argument of as.Date(), you can convert any date format into a Date object. For example, to convert "5 Aug 2016" into a date, use the following:

```
> as.Date("5 Aug 2016", format = "%d %b %Y")
[1] "2016-08-05"
```

This rather cryptic line of code indicates that the date format consists of the day (%d), abbreviated month name (%b), and the year with century (%Y), with spaces between each component.

Table 6-2 lists some of the many date format codes that you can use to specify dates. You can access the full list by typing ?strptime in your R console.

Table 6-2	Some Format Codes for Dates (For Use with as.Date, POSXct, POSIXlt, and strptime)
Format	*Description*
%Y	Year with century.
%y	Year without century (00–99). Values 00 to 68 are prefixed by 20, and values 69 to 99 are prefixed by 19.
%m	Month as decimal number (01–12).
%B	Full month name in the current locale. (Also matches abbreviated name on input.)
%b	Abbreviated month name in the current locale. (Also matches full name on input.)
%d	Day of the month as a decimal number (01–31). You don't need to add the leading zero when converting text to Date, but when you format a Date as text, R adds the leading zero.
%A	Full weekday name in the current locale. (Also matches abbreviated name on input.)
%a	Abbreviated weekday name in the current locale. (Also matches full name on input.)
%w	Weekday as decimal number (0–6, with Sunday being 0).

Try the formatting codes with another common date format, "05/8/2016" (that is, day, month, and year separated by a slash):

```
> as.Date("05/8/2016", format = "%d/%m/%Y")
[1] "2016-08-05"
```

Adding Time Information to Dates

Often, referring only to dates isn't enough. You also need to indicate a specific time in hours and minutes.

To specify time information in addition to dates, you can choose between two functions in R: as.POSIXct() and as.POSIXlt(). These two datetime functions differ in the way that they store date information internally, as well as in the way that you can extract date and time elements. (For more on these two functions, see the nearby sidebar, "The two datetime functions.")

POSIX is the name of a set of standards that refers to the UNIX operating system. In R, you find two types of date-time objects that use this set of standards. These types are called POSIXct and POSIXlt, and you can create and change them using functions with the same name. POSIXct refers to a time that is internally stored as the number of seconds since the start of 1970, by default. (You can modify the origin year by setting the origin argument to POSIXct().) POSIXlt refers to a date stored as a named list of vectors for the year, month, day, hours, and minutes.

According to Wikipedia, the time of the *Apollo 11* moon landing was July 20, 1969, at 20:17:39 UTC. (UTC is the acronym for Coordinated Universal Time. It's how the world's clocks are regulated.) To express this date and time in R, try:

```
> apollo <- "July 20, 1969, 20:17:39"
> apollo.fmt <- "%B %d, %Y, %H:%M:%S"
> xct <- as.POSIXct(apollo, format = apollo.fmt, tz = "UTC")
> xct
[1] "1969-07-20 20:17:39 UTC"
```

As you can see, as.POSIXct() takes similar arguments to as.Date(), but you need to specify the date format as well as the time zone.

Table 6-3 lists additional formatting codes that are useful when working with time information in dates.

Table 6-3	Formatting Codes for the Time Element of POSIXct and POSIXlt Datetimes
Format	**Description**
%H	Hours as a decimal number (00–23)
%I	Hours as a decimal number (01–12)
%M	Minutes as a decimal number (00–59)
%S	Seconds as a decimal number (00–61)
%p	AM/PM indicator

The POSIXct format allows fractional second resolution. This means you can work with time stamps in fractions of a second, such as millisecond, in your date objects.

The two datetime functions

In most computer languages and systems, dates are represented by numeric values that indicate the number of seconds since a specific instant in time (known as the *epoch*).

In R, you can use two functions to work with datetime objects: POSIXct() and POSIXlt(). These functions create objects of class POSIXct and POSIXlt, respectively:

✔ POSIXct objects represent the (signed) number of seconds since the beginning of 1970 (in the UTC time zone) as a numeric vector.

✔ POSIXlt objects are named lists of vectors representing nine elements of a datetime (sec, min, hour, and so on).

Because POSIXct are numbers, and POSIXlt objects are lists, POSIXct objects require less memory.

The following table summarizes the main differences between the different datetime classes in R.

Class	Description	Useful Functions
Date	Calendar date	as.Date()
POSIXct	The number of seconds since the beginning of 1970 (in the UTC time zone) as a numeric vector	as.POSIXct()
POSIXlt	A named list of vectors representing nine elements (sec, min, hour, and so on)	as.POSIXlt()

Formatting Dates and Times

To format a date for pretty printing, you use `format()`, which takes a `POSIXct` or `POSIXlt` datetime as input, together with a formatting string. You have already encountered a formatting string when creating a date.

Continuing with the example where the object `xct` is the day and time of the *Apollo* landing, you can format this date and time in many different ways. For example, to format it as `DD/MM/YY`, try:

```
> format(xct, "%d/%m/%y")
[1] "20/07/69"
```

In addition to the formatting codes, you can use any other character. If you want to format the `xct` datetime as a sentence, try the following:

```
> format(xct, "%M minutes past %I %p, on %d %B %Y")
[1] "17 minutes past 08 PM, on 20 July 1969"
```

You can find the formatting codes in Table 6-2 and Table 6-3, as well as at the Help page `?strptime`.

Performing Operations on Dates and Times

Because R stores datetime objects as numbers, you can do various operations on dates, including addition, subtraction, comparison, and extraction.

Addition and subtraction

R stores objects of class `POSIXct` as the number of seconds since the epoch (usually the start of 1970), so you can do addition and subtraction by adding or subtracting seconds. It's more common to add or subtract days from dates, so it's useful to know that each day has 86,400 seconds.

To add seven days to the *Apollo* landing date, use addition, just remember to multiply the number of days by the number of seconds per day:

```
> xct + 7*86400
[1] "1969-07-27 20:17:39 UTC"
```

Once you know that you can convert any duration to seconds, you can add or subtract any value to a datetime object. For example, add three hours to the time of the *Apollo* moon landing:

```
> xct + 3*60*60
[1] "1969-07-20 23:17:39 UTC"
```

Similarly, to get a date seven days earlier, use subtraction:

```
> xct - 7*86400
[1] "1969-07-13 20:17:39 UTC"
```

There is an important difference between Date objects and POSIXct or POSIXlt objects. If you use a Date object, you add and subtract *days;* with POSIXct and POSIXlt, the operations add or subtract only *seconds.*

Try that yourself, first converting xct to a Date object, then subtracting 7:

```
> as.Date(xct) - 7
[1] "1969-07-13"
```

Comparison of dates

Similar to the way that you can add or subtract dates you can also compare dates with the comparison operators, such as less than (<) or greater than (>), covered in Chapter 4.

Say you want to compare the current time with any fixed time. In R, you use the Sys.time() function to get the current system time:

```
> Sys.time()
[1] "2015-01-16 14:19:56 GMT"
```

Now you know the exact time when we wrote this sentence. Clearly when you try the same command you will get a different result!

Now you can compare your current system time with the time of the *Apollo* landing:

```
> Sys.time() < xct
[1] FALSE
```

If your system clock is accurate, then obviously you would expect the result to be false, because the moon landing happened more than 40 years ago.

As we cover in Chapter 4, the comparison operators are vectorized, so you can compare an entire vector of dates with the moon landing date. Try to use all your knowledge of dates, sequences of dates, and comparison operators to compare the start of several decades to the moon landing date.

Start by creating a POSIXct object containing the first day of 1950. Then use seq() to create a sequence with intervals of ten years:

```
> dec.start <- as.POSIXct("1950-01-01")
> dec <- seq(dec.start, by = "10 years", length.out = 4)
> dec
[1] "1950-01-01 GMT" "1960-01-01 GMT" "1970-01-01 GMT"
[4] "1980-01-01 GMT"
```

Finally, you can compare your new vector dec with the moon landing date:

```
> dec > xct
[1] FALSE FALSE  TRUE  TRUE
```

As you can see, the first two results (comparing the date of the moon landing to 1950 and 1960) are FALSE, and the last two values (comparing the date of the moon landing to 1970 and 1980) are TRUE.

Extraction

Another thing you may want to do is to extract specific elements of the date, such as the day, month, or year. For example, scientists may want to compare the weather in a specific month (say, January) for many different years. To do this, they first have to determine the month, by extracting the months from the datetime object.

An easy way to achieve this is to work with dates in the POSIXlt class, because this type of data is stored internally as a named list, which enables you to extract components by name. To do this, first convert the Date class:

```
> xlt <- as.POSIXlt(xct)
> xlt
[1] "1969-07-20 20:17:39 UTC"
```

Next, use the $ operator to extract the different components. For example, to get the year, use the following:

```
> xlt$year
[1] 69
```

And to get the month, use the following:

```
> xlt$mon
[1] 6
```

You might be surprised to find out that R considers July to be the sixth month of the year. POSIXlt is based on an old Unix format, and that format starts to count from 0 instead of 1. In other words, 0 represents January. Get more information on the exact coding of all values on the help page ?POSIXlt.

You can use the unclass() function to expose the internal structure of POSIXlt objects.

```
> unclass(xlt)
```

If you run this line of code, you'll see that POSIXlt objects are really just named lists. You get to work with lists in much more detail in Chapter 7.

More date and time fun(ctionality)

In this chapter, we barely scratch the surface of how to handle dates and times in R. You may want to explore additional functionality available in R and add-on packages by looking at the following:

✔ chron: In addition to all the data classes that we cover in this chapter, R has the simpler chron class for datetime objects that don't have a time zone. To investigate this class, first load the chron package with library("chron") and then read the Help file ?chron.

✔ lubridate: You can download the add-on package lubridate from CRAN. This package provides many functions to make it easier to work with dates. You can download and find more information at http://cran.r-project.org/web/packages/lubridate/index.html.

R also has very good support for objects that represent time series data. Time series data usually refers to information that was recorded at fixed intervals, such as days, months, or years:

✔ ts: In R, you use the ts() function to create time series objects. These are vector or matrix objects that contain information about the observations, together with information about the start, frequency, and end of each observation period. With ts class data you can use powerful R functions to do modeling and forecasting — for example, arima() is a general model for time series data.

✔ zoo **and** xts: The add-on package zoo extends time series objects by allowing observations that don't have such strictly fixed intervals. You can download it from CRAN at http://cran.r-project.org/web/packages/zoo/index.html. The add-on package xts provides additional extensions to time series data and builds on the functionality of ts as well as zoo objects. You can download xts from CRAN: http://cran.r--project.org/web/packages/xts/index.html.

Now you have all the information to go on a date with R and enjoy the experience!

Chapter 7

Working in More Dimensions

*I*n the previous chapters, you worked with one-dimensional vectors. The data could be represented by a single row or column in a Microsoft Excel spreadsheet. But often you need more than one dimension. Many calculations in statistics are based on matrices, so you need to be able to represent matrices and perform matrix calculations. Many datasets contain values of different types for multiple variables and observations, so you need a two-dimensional table to represent this data. In Excel, you would do that in a spreadsheet; in R, you use a specific object called a *data frame* for the task.

Adding a Second Dimension

In the previous chapters, you constructed vectors to hold data in a one-dimensional structure. In addition to vectors, R can represent matrices as an object you work and calculate with. In fact, R really shines when it comes to matrix calculations and operations. In this section, we take a closer look at the magic you can do with them.

Discovering a new dimension

Vectors are closely related to a bigger class of objects, arrays. Arrays have two very important features:

✔ They contain only a single type of value.

✔ They have dimensions.

The dimensions of an array determine the type of the array. You know already that a vector has only one dimension. An array with two dimensions is a *matrix*. Anything with more than two dimensions is simply called an *array*. You find a graphical representation of this in Figure 7-1.

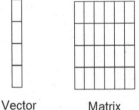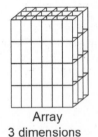

Figure 7-1:
A vector, a matrix, and an array.

Vector — 1 dimension   Matrix — 2 dimensions   Array — 3 dimensions

Technically, a vector has no dimensions at all in R. R returns NULL as a result if you use the functions dim(), nrow(), or ncol() (mentioned in the section "Looking at the properties" later in this chapter) with a vector as argument.

Creating your first matrix

Creating a matrix is almost as easy as writing the word: You simply use the matrix() function. You do have to give R a little bit more information, though. R needs to know which values you want to put in the matrix and how you want to put them in. The matrix() function has several arguments for this:

✔ data is a vector of values you want in the matrix.

✔ ncol takes a single number that tells R how many columns you want.

✔ nrow takes a single number that tells R how many rows you want.

✔ byrow takes a logical value that tells R whether you want to fill the matrix row-wise (TRUE) or column-wise (FALSE). Column-wise is the default.

So, the following code results in a matrix with the numbers 1 through 12, in four columns and three rows.

```
> first.matrix <- matrix(1:12, ncol = 4)
> first.matrix
     [,1] [,2] [,3] [,4]
[1,]    1    4    7   10
[2,]    2    5    8   11
[3,]    3    6    9   12
```

You don't have to specify both `ncol` and `nrow`. If you specify one, R will know automatically what the other needs to be.

Alternatively, if you want to fill the matrix row by row, try:

```
> matrix(1:12, ncol = 4, byrow = TRUE)
     [,1] [,2] [,3] [,4]
[1,]    1    2    3    4
[2,]    5    6    7    8
[3,]    9   10   11   12
```

Looking at the properties

You can look at the structure of an object using the `str()` function. If you do that for your first matrix, you get the following result:

```
> str(first.matrix)
 int [1:3, 1:4] 1 2 3 4 5 6 7 8 9 10 ...
```

This looks remarkably similar to the output for a vector, with the difference that R displays both the indices for the rows and for the columns. If you want the number of rows and columns without looking at the structure, you can use the `dim()` function.

```
> dim(first.matrix)
[1] 3 4
```

To get only the number of rows, you use the `nrow()` function. The `ncol()` function gives you the number of columns of a matrix.

You can find the total number of values in a matrix exactly the same way as you do with a vector, using the `length()` function:

```
> length(first.matrix)
[1] 12
```

Actually, if you look at the output of the str() function, that matrix looks very much like a vector. That's because, internally, it's a vector with a small extra piece of information that tells R the dimensions (see the nearby sidebar, "Playing with attributes"). You can use this property of matrices in calculations, as you'll see further in this chapter.

Playing with attributes

Both the names and the dimensions of matrices and arrays are stored in R as *attributes* of the object. These attributes can be seen as labeled values you can attach to any object. They form one of the mechanisms R uses to define specific object types like dates, time series, and so on. They can include any kind of information, and you can use them yourself to add information to any object.

To illustrate, re-create my.array and baskets.team (created in the later sections "Adding more dimensions" and "Combining vectors into a matrix").

```
> my.array <- array(1:24, dim =
       c(3, 4, 2))
> baskets.team <- rbind(
+    baskets.of.Granny =
         c(12, 4, 5, 6, 9, 3),
+    baskets.of.Geraldine =
         c(5, 4, 2, 4, 12, 9)
+ )
```

To see all the attributes of an object, use the attributes() function. To see all the attributes of my.array, try:

```
> attributes(my.array)
$dim
[1] 3 4 2
```

This function returns a named list, where each item in the list is an attribute. Each attribute can, itself, be a list again. For example, the attribute dimnames is actually a list containing the row and column names of a matrix. You can check that for yourself by checking the output of attributes(baskets.team). You can also set all attributes as a named list — find examples in the Help file ?attributes.

To get or set a single attribute, use the attr() function. This function takes two important arguments. The first argument is the object you want to examine, and the second argument is the name of the attribute you want to see or change. If the attribute you ask for doesn't exist, R simply returns NULL.

Imagine you want to add which season Granny and Geraldine scored the baskets mentioned in baskets.team. You can do this with:

```
> attr(baskets.team, "season")
       <- "2010-2011"
```

To get the value of this attribute returned, use:

```
> attr(baskets.team, "season")
[1] "2010-2011"
```

You can remove attributes by setting their value to NULL:

```
> attr(baskets.team, "season") <- NULL
```

Combining vectors into a matrix

In Chapter 4, you create two vectors that contain the number of baskets Granny and Geraldine made in the six games of this basketball season. It would be nicer, though, if the number of baskets for the whole team were contained in one object. With matrices, this becomes possible. You can combine both vectors as two rows of a matrix with the rbind() function:

```
> baskets.of.Granny <- c(12, 4, 5, 6, 9, 3)
> baskets.of.Geraldine <- c(5, 4, 2, 4, 12, 9)
> baskets.team <- rbind(baskets.of.Granny, baskets.of.Geraldine)
```

The object baskets.team is a matrix, and the rows take the names of the original vectors. You work with these names in the next section.

```
> baskets.team
                    [,1] [,2] [,3] [,4] [,5] [,6]
baskets.of.Granny     12    4    5    6    9    3
baskets.of.Geraldine   5    4    2    4   12    9
```

The cbind() function does something similar. It binds the vectors as columns of a matrix:

```
> cbind(1:3, 4:6, matrix(7:12, ncol = 2))
     [,1] [,2] [,3] [,4]
[1,]    1    4    7   10
[2,]    2    5    8   11
[3,]    3    6    9   12
```

Here you bind together three different nameless objects:

✔ A vector with the values 1 to 3 (1:3)

✔ A vector with the values 4 to 6 (4:6)

✔ A matrix with two columns and three rows, filled column-wise with the values 7 through 12 (matrix(7:12, ncol = 2))

This example shows some other properties of cbind() and rbind() that can be very useful:

✔ The functions work with both vectors and matrices. They also work on other objects, as shown in the "Manipulating Values in a Data Frame" section, later in this chapter.

✔ You can give more than two arguments to either function. The objects are combined in the order you specify them in the arguments.

✔ You can combine different types of objects, as long as the dimensions fit. Here you combine vectors and matrices in one function call.

Using the Indices

If you look at the output of the code in the previous section, you'll probably notice the brackets you used in the previous chapters for accessing values in vectors through the indices. But this time, these indices look a bit different. Whereas a vector has only one dimension that can be indexed, a matrix has two. You separate the indices for both dimensions by a comma — you give the index for the row before the comma, and the index for the column after the comma.

Extracting values from a matrix

You can use these indices the same way you use vectors in Chapter 4. You can assign and extract values, use numerical or logical indices, drop values by using a minus sign, and so forth.

Using numeric indices

For example, to extract the values in the first two rows and the last two columns, try:

```
> first.matrix[1:2, 2:3]
     [,1] [,2]
[1,]    4    7
[2,]    5    8
```

R returns a matrix. Pay attention to the indices of this new matrix — they're not the indices of the original matrix anymore.

R gives you an easy way to extract *complete* rows and columns from a matrix. You simply don't specify the other dimension. For example, to get the second and third rows from your first matrix, try:

```
> first.matrix[2:3, ]
     [,1] [,2] [,3] [,4]
[1,]    2    5    8   11
[2,]    3    6    9   12
```

Dropping values using negative indices

In Chapter 4, you drop values in a vector by using a negative value for the index. This little trick works perfectly well with matrices, too. So, to get all the values except the second row and third column of first.matrix, try:

```
> first.matrix[-2, -3]
     [,1] [,2] [,3]
[1,]    1    4   10
[2,]    3    6   12
```

With matrices, a negative index always means: "Drop the complete row or column." If you want to drop only the element at the second row and the third column, you have to treat the matrix like a vector. So, in this case, you drop the second element in the third column like this:

```
> nr <- nrow(first.matrix)
> id <- nr * 2 + 2
> first.matrix[-id]
 [1]  1  2  3  4  5  6  7  9 10 11 12
```

This returns a vector, because the 11 remaining elements don't fit into a matrix anymore. Now what happened here exactly? Remember that matrices are read column-wise. To get the second element in the third column, you need to do the following:

1. **Count the number of rows, using `nrow()`, and store that in a variable — for example `nr`.**

   You don't have to do this, but it makes the code easier to read.

2. **Count two columns and then add 2 to get the second element in the third column.**

   Again store this result in a variable (for example, `id`).

3. **Use the one-dimensional vector extraction `[ ]` to drop this value, as shown in Chapter 4.**

You can do this in one line, like this:

```
> first.matrix[-(2 * nrow(first.matrix) + 2)]
 [1]  1  2  3  4  5  6  7  9 10 11 12
```

This is just one example of how you can work with indices while treating a matrix like a vector. It requires a bit of thinking at first, but tricks like these can offer very neat solutions to more complex problems as well, especially if you need your code to run as fast as possible.

Juggling dimensions

As with vectors, you can combine multiple numbers in the indices. If you want to drop the first and third rows of the matrix, you can do so like this:

```
> first.matrix[-c(1, 3), ]
 [1]  2  5  8 11
```

Wait a minute. . . . There's only one index. R doesn't return a matrix here — it returns a vector!

By default, R always tries to simplify the objects to the smallest number of dimensions possible when you use the brackets to extract values from an array. So, if you ask for only one column or row, R returns a vector by dropping a dimension.

You can force R to keep all dimensions by using the extra argument `drop` from the indexing function. To get the second row returned as a matrix, you do the following:

```
> first.matrix[2, , drop = FALSE]
     [,1] [,2] [,3] [,4]
[1,]    2    5    8   11
```

This seems like utter magic, but it's not that difficult. You supply three arguments between the brackets, separated by commas. The first argument is the row index. The second argument is the column index. But then what?

Actually, the square brackets work like a function, and the row index and column index are arguments for the square brackets. Now you add an extra argument `drop` with the value `FALSE`. As you do with any other function, you separate the arguments by commas. Put all this together, and you have the code shown here.

The default dropping of dimensions of R can be handy, but it's famous for being overlooked. It can cause serious mishap if you aren't aware of it. Particularly in code where you take a subset of a matrix, you can easily forget about the case where only one row or column is selected.

Replacing values in a matrix

Replacing values in a matrix is done in a very similar way to replacing values in a vector. To replace the value in the second row and third column of `first.matrix` with 4, try:

```
> first.matrix[3, 2] <- 4
> first.matrix
     [,1] [,2] [,3] [,4]
[1,]    1    4    7   10
[2,]    2    5    8   11
[3,]    3    4    9   12
```

You can change an entire row or column of values by not specifying the other dimension. Note that values are recycled, so to change the second row to the sequence 1, 3, 1, 3, try:

```
> first.matrix[2, ] <- c(1, 3)
> first.matrix
     [,1] [,2] [,3] [,4]
[1,]    1    4    7   10
[2,]    1    3    1    3
[3,]    3    4    9   12
```

You can replace a subset of values within the matrix by another matrix. You don't even have to specify the values as a matrix — a vector will do. Try:

```
> first.matrix[1:2, 3:4] <- c(8, 4, 2, 1)
> first.matrix
     [,1] [,2] [,3] [,4]
[1,]    1    4    8    2
[2,]    1    3    4    1
[3,]    3    4    9   12
```

Here you change the values in the first two rows and the last two columns to the numbers 8, 4, 2, and 1.

R reads and writes matrices column-wise by default. So, if you put a vector in a matrix or a subset of a matrix, the values will be added column-wise regardless of the method. If you want to do this row-wise, you first have to construct a matrix with the values using the argument byrow=TRUE. Then you use this matrix instead of the original vector to insert the values.

Naming Matrix Rows and Columns

The rbind() function conveniently added the names of the vectors baskets.of.Granny and baskets.of.Geraldine to the rows of the matrix baskets.team in the previous section. You name the values in a vector in Chapter 5, and you can do something very similar with rows and columns in a matrix.

For that, you have the functions rownames() and colnames(). Guess which one does what? Both functions work much like the names() function you use when naming vector values.

Changing the row and column names

The matrix `baskets.team` from the previous section already has some row names. It would be better if the names of the rows would just read `"Granny"` and `"Geraldine"`. You can easily change these row names like this:

```
> rownames(baskets.team) <- c("Granny", "Geraldine")
```

You can look at the matrix to check if this did what it's supposed to do, or you can take a look at the row names itself like this:

```
> rownames(baskets.team)
[1] "Granny"    "Geraldine"
```

The `colnames()` function works exactly the same. You can, for example, add the number of the game as a column name using the following code:

```
> colnames(baskets.team) <- c("1st", "2nd", "3th", "4th", "5th", "6th")
```

This gives you the following matrix:

```
> baskets.team
          1st 2nd 3th 4th 5th 6th
Granny     12   4   5   6   9   3
Geraldine   5   4   2   4  12   9
```

This is almost like you want it, but the third column name contains an annoying writing mistake. No problem there, R allows you to easily correct that mistake. Just as the with `names()` function, you can use indices to extract or to change a specific row or column name. You can correct the mistake in the column names like this:

```
> colnames(baskets.team)[3] <- "3rd"
```

If you want to get rid of either column names or row names, the only thing you need to do is set their value to `NULL`. This also works for vector names, by the way. You can try that out yourself on a copy of the matrix `baskets.team` like this:

```
> baskets.copy <- baskets.team
> colnames(baskets.copy) <- NULL
> baskets.copy
          [,1] [,2] [,3] [,4] [,5] [,6]
Granny      12    4    5    6    9    3
Geraldine    5    4    2    4   12    9
```

R stores the row and column names in an attribute called `dimnames`. Use the `dimnames()` function to extract or set those values. (See the "Playing with attributes" sidebar, earlier in this chapter, for more information.)

Using names as indices

These row and column names can be used just like you use names for values in a vector, as explained in Chapter 5. You can use these names instead of the index number to select values from a vector. This works for matrices as well, using the row and column names.

Say you want to select the second and the fifth game for both ladies, try:

```
> baskets.team[, c("2nd", "5th")]
          2nd 5th
Granny      4   9
Geraldine   4  12
```

Exactly as before, you get all rows if you don't specify which ones you want. Alternatively, you can extract all the results for Granny like this:

```
> baskets.team["Granny", ]
1st 2nd 3rd 4th 5th 6th
 12   4   5   6   9   3
```

That's the result, indeed, but the row name is gone now. As explained in the "Juggling dimensions" section, earlier in this chapter, R tries to simplify the matrix to a vector, if that's possible. In this case, a single row is returned so, by default, this result is transformed to a vector.

If a one-row matrix is simplified to a vector, the column names are used as names for the values. If a one-column matrix is simplified to a vector, the row names are used as names for the vector. If you want to keep all names, you must set the argument `drop` to `FALSE` to avoid conversion to a vector.

Calculating with Matrices

Probably the strongest feature of R is the ease of dealing with matrix operations in an easy and optimized way. Because much of statistics boils down to matrix operations, it's only natural that R loves to crunch those numbers.

Using standard operations with matrices

When talking about operations on matrices, you can treat either the elements of the matrix or the whole matrix as the value you operate on. That difference is pretty clear when you compare, for example, transposing a matrix and adding a single number (or *scalar*) to a matrix. When transposing, you work with the whole matrix. When adding a scalar to a matrix, you add that scalar to every element of the matrix.

You add a scalar to a matrix simply by using the addition operator, +, like this:

```
> first.matrix + 4
     [,1] [,2] [,3] [,4]
[1,]    5    8   12    6
[2,]    5    7    8    5
[3,]    7    8   13   16
```

You can use all other arithmetic operators in exactly the same way to perform an operation on all elements of a matrix.

The difference between operations on matrices and elements becomes less clear if you talk about adding matrices together. In fact, the addition of two matrices is the addition of their corresponding elements. So, you need to make sure both matrices have the same dimensions.

Look at another example: Say you want to add 1 to the first row, 2 to the second row, and 3 to the third row of the matrix first.matrix. You can do this by constructing a matrix second.matrix that has four columns and three rows and that has 1, 2, and 3 as values in the first, second, and third rows, respectively. Using the recycling of the first argument by the matrix function (see Chapter 4), you can try:

```
> second.matrix <- matrix(1:3, nrow = 3, ncol = 4)
```

With the addition operator, you can add both matrices together, like this:

```
> first.matrix + second.matrix
     [,1] [,2] [,3] [,4]
[1,]    2    5    9    3
[2,]    3    5    6    3
[3,]    6    7   12   15
```

This is the solution your math teacher would approve of if she asked you to do the matrix addition of the first and second matrix. And even more, if the dimensions of both matrices are not the same, R will complain and refuse to carry out the operation, as you can see:

```
> first.matrix + second.matrix[, 1:3]
Error in first.matrix + second.matrix[, 1:3] : non-conformable arrays
```

But what would happen if instead of adding a matrix, we added a vector? Try this:

```
> first.matrix + 1:3
     [,1] [,2] [,3] [,4]
[1,]    2    5    9    3
[2,]    3    5    6    3
[3,]    6    7   12   15
```

R does not complain about the dimensions, and recycles the vector over the values of the matrices! In fact, R treats the matrix as a vector by simply ignoring the dimensions. So, in this case, you don't use matrix addition but simple (vectorized) addition (see Chapter 4).

By default, R fills matrices column-wise. Whenever R reads a matrix, it also reads it column-wise. This has important implications for the work with matrices. If you don't stay aware of this, R can bite you in the leg nastily.

Calculating row and column summaries

In Chapter 4, you summarize vectors using functions like sum() and prod(). All these functions work on matrices as well, because a matrix is simply a vector with dimensions attached to it. You also can summarize the rows or columns of a matrix using some specialized functions.

In the previous section, you created a matrix baskets.team with the number of baskets that both Granny and Geraldine made in the previous basketball season. To get the total number each woman made during the last six games, use the function rowSums():

```
> rowSums(baskets.team)
  Granny Geraldine
      39        36
```

The rowSums() function returns a named vector with the sums of each row.

You can get the means of each row with rowMeans(), and the respective sums and means of each column with colSums() and colMeans().

Doing matrix arithmetic

Apart from the classical arithmetic operators, R contains a large set of operators and functions to perform a wide set of matrix operations. Many of these operations are used in advanced mathematics, so you may never need them. Some of them can come in pretty handy, though, if you need to flip around data or you want to calculate some statistics yourself.

Transposing a matrix

Flipping around a matrix so the rows become columns and vice versa is very easy in R. Use the t() function (which stands for *transpose*):

```
> t(first.matrix)
     [,1] [,2] [,3]
[1,]    1    2    3
[2,]    4    5    6
[3,]    7    8    9
[4,]   10   11   12
```

You can also try this with a vector. Since matrices are read and filled column-wise, it shouldn't come as a surprise that the t() function sees a vector as a one-column matrix. Thus the transpose of a vector is a single-row matrix:

```
> t(1:10)
     [,1] [,2] [,3] [,4] [,5] [,6] [,7] [,8] [,9] [,10]
[1,]    1    2    3    4    5    6    7    8    9    10
```

You can tell this is a matrix by the dimensions. This seems trivial, but imagine you're selecting only one row from a matrix and transposing it. Unlike what you may expect, you get a row instead of a column:

```
> t(first.matrix[2, ])
     [,1] [,2] [,3] [,4]
[1,]    2    5    8   11
```

Inverting a matrix

Contrary to your intuition, inverting a matrix is not done by raising it to the power of –1. As explained in Chapter 6, R normally applies the arithmetic operators element-wise on the matrix. So, the command first.matrix^(-1) doesn't give you the inverse of the matrix; instead, it gives you the inverse of the elements. To invert a matrix, you use the solve() function, like this:

```
> square.matrix <- matrix(c(1, 0, 3, 2, 2, 4, 3, 2, 1), ncol = 3)
> solve(square.matrix)
      [,1]        [,2]        [,3]
[1,]  0.5 -0.8333333   0.1666667
[2,] -0.5  0.6666667   0.1666667
[3,]  0.5 -0.1666667  -0.1666667
```

Be careful inverting a matrix like this because of the risk of round-off errors. R computes most statistics based on decompositions like the QR decomposition, single-value decomposition, and Cholesky decomposition. You can do that yourself using the functions `qr()`, `svd()`, and `chol()`, respectively. Check the respective Help pages for more information.

Multiplying two matrices

The multiplication operator (*) works element-wise on matrices. To calculate the inner product of two matrices, you use the special operator %*%:

```
> first.matrix %*% t(second.matrix)
     [,1] [,2] [,3]
[1,]   22   44   66
[2,]   26   52   78
[3,]   30   60   90
```

You have to transpose the `second.matrix` first; otherwise, both matrices have non-conformable dimensions. Multiplying a matrix with a vector is a bit of a special case; as long as the dimensions fit, R automatically converts the vector to either a row or a column matrix, whatever is applicable in that case. You can check for yourself in the following example:

```
> first.matrix %*% 1:4
     [,1]
[1,]   70
[2,]   80
[3,]   90
> 1:3 %*% first.matrix
     [,1] [,2] [,3] [,4]
[1,]   14   32   50   68
```

Adding More Dimensions

Matrices are special cases of a more general type of object, arrays. All arrays can be seen as a vector with an extra dimension attribute, and the number of dimensions is completely arbitrary. This also means that you can construct an array with only one dimension. Although technically this is not the same as a vector, in most cases it will behave exactly the same.

Although arrays with more than two dimensions are not often used in R, it's good to know of their existence. They can be useful in certain cases, like when you want to represent two-dimensional data in a time series or store multi-way tables in R.

Creating an array

You have two different options for constructing matrices or arrays. Either you use the creator functions `matrix()` and `array()`, or you simply change the dimensions using the `dim()` function.

Using the creator functions

You can create an array easily with the `array()` function, where you give the data as the first argument and a vector with the sizes of the dimensions as the second argument. The number of dimension sizes in that argument gives you the number of dimensions. For example, you make an array with four columns, three rows, and two "slices" like this:

```
> my.array <- array(1:24, dim = c(3, 4, 2))
> my.array
, , 1

     [,1] [,2] [,3] [,4]
[1,]    1    4    7   10
[2,]    2    5    8   11
[3,]    3    6    9   12


, , 2

     [,1] [,2] [,3] [,4]
[1,]   13   16   19   22
[2,]   14   17   20   23
[3,]   15   18   21   24
```

This array has three dimensions. Notice that, although the rows are given as the first dimension, the slices are filled column-wise. So, for arrays, R fills the columns, then the rows, and then the rest.

Changing the dimensions of a vector

Alternatively, you could just add the dimensions using the `dim()` function. This is a little hack that goes a bit faster than using the `array()` function; it's especially useful if you have your data already in a vector. (This little trick also works for creating matrices, by the way, because a matrix is nothing more than an array with only two dimensions.)

Say you already have a vector with the numbers 1 through 24, like this:

```
> my.vector <- 1:24
```

You can easily convert that vector to an array exactly like my.array simply by assigning the dimensions, like this:

```
> dim(my.vector) <- c(3, 4, 2)
```

If you check how my.vector looks like now, you see there is no difference from the array my.array that you created before.

You can check whether two objects are identical by using the identical() function. To check whether my.vector and my.array are identical, try:

```
> identical(my.array, my.vector)
[1] TRUE
```

Using dimensions to extract values

Extracting values from an array with any number of dimensions is completely equivalent to extracting values from a matrix. You separate the dimension indices you want to retrieve with commas, and if necessary you can use the drop argument exactly as you do with matrices. For example, to get the value from the second row and third column of the first slice of my.array, try:

```
> my.array[2, 3, 1]
[1] 8
```

If you want the third column of the second slice as an array, use:

```
> my.array[, 3, 2, drop = FALSE]
, , 1

      [,1]
[1,]   19
[2,]   20
[3,]   21
```

If you don't specify the drop=FALSE argument, R will try to simplify the object as much as possible. This also means that if the result has only two dimensions, R will make it a matrix. To return a matrix that consists of the second row of each slice, use:

```
> my.array[2, , ]
      [,1] [,2]
[1,]    2   14
[2,]    5   17
[3,]    8   20
[4,]   11   23
```

This reduction doesn't mean, however, that rows stay rows. In this case, R made the rows columns. This is due to the fact that R first selects the values, and then adds the dimensions necessary to represent the data correctly. In this case R needs two dimensions with four indices (the number of columns) and two indices (the number of slices), respectively. As R fills a matrix column-wise, the original rows now turned into columns.

Combining Different Types of Values in a Data Frame

Until this point in the book, you combine values of the same type into either a vector or a matrix. But datasets are, in general, built up from different data types. You can have, for example, the names of your employees, their salaries, and the date they started at your company all in the same dataset. But you can't combine all this data in one matrix without converting the data to character data. So, you need a new data structure to keep all this information together in R. That data structure is a *data frame*.

Creating a data frame from a matrix

Take a look again at the number of baskets scored by Granny and her friend Geraldine. In the "Adding a second dimension" section, earlier in this chapter, you created a matrix baskets.team with the number of baskets for both ladies. It makes sense to make this matrix a data frame with two variables: one for Granny's baskets and one for Geraldine's baskets.

Using the function as.data.frame

To convert the matrix baskets.team into a data frame, you use the function as.data.frame():

```
> baskets.df <- as.data.frame(t(baskets.team))
```

You don't have to use the transpose function, t(), to create a data frame, but in our example we want each player to be a separate variable. With data frames, each variable is a column, but in the original matrix, the rows represent the baskets for a single player. So, in order to get the desired result, you first have to transpose the matrix with t() before converting the matrix to a data frame with as.data.frame().

Looking at the structure of a data frame

If you take a look at the object, it looks exactly the same as the transposed matrix t(baskets.team):

```
> baskets.df
    Granny Geraldine
1st     12        5
2nd      4        4
3rd      5        2
4th      6        4
5th      9       12
6th      3        9
```

But there is a very important difference between the two: baskets.df is a data frame. This becomes clear if you take a look at the internal structure of the object, using the str() function:

```
> str(baskets.df)
'data.frame': 6 obs. of  2 variables:
 $ Granny   : num  12 4 5 6 9 3
 $ Geraldine: num  5 4 2 4 12 9
```

Now this starts looking more like a real dataset. You can see in the output that you have six observations and two variables. The variables are called Granny and Geraldine. It's important to realize that each variable in itself is a vector; hence, it has one of the types you learn about in Chapters 4, 5, and 6. In this case, the output tells you that both variables are numeric.

Counting values and variables

To know how many observations a data frame has, you can use the nrow() function as you would with a matrix, like this:

```
> nrow(baskets.df)
[1] 6
```

Likewise, the ncol() function gives you the number of variables. But you can also use the length() function to get the number of variables for a data frame:

```
> length(baskets.df)
[1] 2
```

Creating a data frame from scratch

The conversion from a matrix to a data frame can't be used to construct a data frame with different types of values. If you combine both numeric and character data in a matrix, for example, everything will be converted to character. You can construct a data frame from scratch, though, using the `data.frame()` function.

Making a data frame from vectors

So, let's make a little data frame with the names, salaries, and starting dates of a few imaginary co-workers. First, you create three vectors that contain the necessary information like this:

```
> employee <- c("John Doe", "Peter Gynn", "Jolie Hope")
> salary <- c(21000, 23400, 26800)
> startdate <- as.Date(c("2010-11-1", "2008-3-25", "2007-3-14"))
```

Now you have three different vectors in your workspace:

- ✔ A **character vector** called `employee`, containing the names
- ✔ A **numeric vector** called `salary`, containing the yearly salaries
- ✔ A **date vector** called `startdate`, containing the dates on which the contracts started

Next, you combine the three vectors into a data frame:

```
> employ.data <- data.frame(employee, salary, startdate)
```

The result is a data frame, `employ.data`, with the following structure:

```
> str(employ.data)
'data.frame': 3 obs. of  3 variables:
 $ employee : Factor w/ 3 levels "John Doe","Jolie Hope",..: 1 3 2
 $ salary   : num  21000 23400 26800
 $ startdate: Date, format: "2010-11-01" "2008-03-25" ...
```

To combine a number of vectors into a data frame, you simple add all vectors as arguments to the `data.frame()` function, separated by commas. R will create a data frame with variables that are named the same as the vectors used. Keep in mind that these vectors must have the same length.

Keeping characters as characters

You may have noticed something odd when looking at the structure of `employ.data`. Whereas the vector `employee` is a character vector, R made the variable `employee` in the data frame a factor.

R does this by default, but you have an extra argument to the `data.frame()` function that can avoid this — namely, the argument `stringsAsFactors`. In the `employ.data` example, you can prevent the transformation to a factor of the `employee` variable by using the following code:

```
> employ.data <- data.frame(employee, salary, startdate,
+                           stringsAsFactors = FALSE)
```

If you look at the structure of the data frame now, you see that the variable `employee` is a character vector, as shown in the following output:

```
> str(employ.data)
'data.frame': 3 obs. of  3 variables:
 $ employee : chr  "John Doe" "Peter Gynn" "Jolie Hope"
 $ salary   : num  21000 23400 26800
 $ startdate: Date, format: "2010-11-01" "2008-03-25" ...
```

By default, R always transforms character vectors to factors when creating a data frame with character vectors or converting a character matrix to a data frame. This can be a nasty cause of errors. If you make it a habit to always specify the `stringsAsFactors` argument, you can always deduce from your code whether or not the conversion to factors happened.

Naming variables and observations

If you look at the data frame `baskets.df` you created in the preceding section, you see something similar to the column and row names of a matrix. R allows you to name both the variables and the observations in a dataset. You can actually use these names in the same way as you use row and column names in a matrix, but there are a few differences as well. We discuss these next.

Working with variable names

Variables in a data frame always have a name. Even if you didn't specify them yourself, R will always try to give variables a sensible name. To access the variable names, you can again treat a data frame like a matrix and use the function `colnames()` like this:

```
> colnames(employ.data)
[1] "employee" "salary"   "startdate"
```

But, in fact, this is taking the long way around. In case of a data frame, the `colnames()` function lets the hard work be done internally by another function, the `names()` function. So, to get the variable names, you can just use that function directly like this:

```
> names(employ.data)
[1] "employee" "salary"   "startdate"
```

You can use that same function to assign new names to the variables as well. For example, to rename the variable `startdate` to `firstday`, you can use the following code:

```
> names(employ.data)[3] <- "firstday"
> names(employ.data)
[1] "employee" "salary"  "firstday"
```

Naming observations

One important difference between a matrix and a data frame is that data frames always have named observations. Whereas the `rownames()` function returns `NULL` if you didn't specify the row names of a matrix, it will always give a result in the case of a data frame.

Check the outcome of the following code:

```
> rownames(employ.data)
[1] "1" "2" "3"
```

By default, the row names — or observation names — of a data frame are simply the row numbers in character format. You can't get rid of them, even if you try to delete them by assigning the `NULL` value (as you can do with matrices).

You shouldn't try to get rid of them either, because your data frame won't be displayed correctly any more if you do.

You can, however, change the row names exactly as you do with matrices, simply by assigning the values via the `rownames()` function, like this:

```
> rownames(employ.data) <- c("Chef", "BigChef", "BiggerChef")
> employ.data
            employee salary    firstday
Chef         John Doe  21000  2010-11-01
BigChef     Peter Gynn 23400  2008-03-25
BiggerChef  Jolie Hope 26800  2007-03-14
```

Don't be fooled, though: Row names can look like another variable, but you can't access them the way you access the variables.

Manipulating Values in a Data Frame

Creating a data frame is nice, but data frames would be pretty useless if you couldn't change the values or add data to them. Luckily, data frames have a very nice feature: When it comes to manipulating the values, almost all tricks you use on matrices also can be used on data frames. You can also use some methods that are designed specifically for data frames. In this next section, we explain these methods. We use the data frame `baskets.df` that you created in the "Creating a data frame from a matrix" section, earlier in this chapter.

Extracting variables, observations, and values

In many cases, you can extract values from a data frame by pretending that it's a matrix. But although data frames may look like matrices, they definitely are not. Unlike matrices and arrays, data frames are not internally stored as vectors but as *lists of vectors*. You start with lists in the "Combining different objects in a list" section, later in this chapter.

Pretending it's a matrix

If you want to extract values from a data frame, you can just pretend it's a matrix and start from there. You can use index numbers, names, or logical vectors for selection, like you would with matrices. For example, you can get the number of baskets scored by Geraldine in the third game like this:

```
> baskets.df["3rd", "Geraldine"]
[1] 2
```

Likewise, you can get all the baskets that Granny scored using the column index, like this:

```
> baskets.df[, 1]
[1] 12  4  5  6  9  3
```

Or, if you want this to be a data frame, you can use the argument `drop=FALSE` exactly as you do with matrices:

```
> str(baskets.df[, 1, drop = FALSE])
'data.frame': 6 obs. of  1 variable:
 $ Granny: num  12 4 5 6 9 3
```

Note that, unlike with matrices, the row names are dropped if you don't specify the `drop=FALSE` argument.

Putting your dollar where your data is

As a careful reader, you noticed already that every variable is preceded by a dollar sign ($) in the output from `str()`. R isn't necessarily pimping your data here — the dollar sign is simply a specific way for accessing variables. To access the variable `Granny`, you can use the dollar sign like this:

```
> baskets.df$Granny
[1] 12  4  5  6  9  3
```

So you specify the data frame, followed by a dollar sign and then the name of the variable. You don't have to surround the variable name by quotation marks (as you would when you use the indices). R will return a vector with all the values contained in that variable. Note again that the row names are dropped here.

With this dollar-sign method, you can access only one variable at a time. If you want to access multiple variables at once using their names, you need to use the square brackets, as in the preceding section.

Adding observations to a data frame

As time goes by, new data may appear and needs to be added to the dataset. Just like matrices, data frames can be appended using the `rbind()` function.

Adding a single observation

Say that Granny and Geraldine played another game with their team, and you want to add the number of baskets they made. The `rbind()` function lets you do that easily:

```
> result <- rbind(baskets.df, c(7, 4))
> result
     Granny Geraldine
1st      12         5
2nd       4         4
3rd       5         2
4th       6         4
5th       9        12
6th       3         9
7         7         4
```

The data frame `result` now has an extra observation compared to `baskets.df`. As explained in the earlier section "Combining vectors into a matrix," `rbind()` can take multiple arguments, as long as they're compatible. In this case, you bind a vector `c(7, 4)` at the bottom of the data frame.

Note that R, by default, sets the row number as the row name for the added rows. You use the `rownames()` function to adjust this, or you can immediately specify the row name between quotes in the `rbind()` function:

```
> baskets.df <- rbind(baskets.df, "7th" = c(7, 4))
```

Note that you must use quotation marks around 7th, because it starts with a number. Without quotation marks, R doesn't recognize it as a name. If you check the object `baskets.df` now, you see the extra observation at the bottom with the correct row name:

```
> baskets.df
    Granny Geraldine
1st     12        5
2nd      4        4
3rd      5        2
4th      6        4
5th      9       12
6th      3        9
7th      7        4
```

Alternatively, you can use indexing to add an extra observation. You see how in the next section.

Adding a series of new observations using rbind

If you need to add multiple new observations to a data frame, doing it one-by-one is not entirely practical. Luckily, you can use `rbind()` to attach a matrix or a data frame with new observations to the original data frame. The matching of the columns is done by name, so you need to make sure that the columns in the matrix or the variables in the data frame with new observations match the variable names in the original data frame.

Let's add another two game results to the data frame `baskets.df`. First, you construct a new data frame with the number of baskets Granny and Geraldine scored, like this:

```
> new.baskets <- data.frame(Granny = c(3, 8), Geraldine = c(9, 4))
```

If you use the `data.frame()` function to construct a new data frame, you can immediately set the variable names by specifying them in the function call, as in the preceding example. That code creates a data frame with the variables Granny and Geraldine where each variable contains the vector given after the equal sign.

To be able to bind the data frame `new.baskets` to the original `baskets.df`, you have to make sure that the variable names match exactly, including the case.

Next, you add the optional row names and the necessary column names with the following code:

```
> rownames(new.baskets) <- c("8th", "9th")
```

To add the matrix to the data frame, you simply do the following:

```
> baskets.df <- rbind(baskets.df, new.baskets)
```

You can try yourself to do the same thing using a data frame instead of a matrix. In Chapter 13, you use more advanced techniques for combining data from different data frames.

Adding a series of values using indices

You also can use the indices to add a set of new observations at one time. You get exactly the same result if you change all the previous code by this simple line:

```
> baskets.df[c("8th", "9th"), ] <- matrix(c(3, 8, 9, 4), ncol = 2)
```

With this code, you do the following:

- ✔ Create a matrix with two columns.
- ✔ Create a vector with the row names 8th and 9th.
- ✔ Use this vector as row indices for the data frame `baskets.df`.
- ✔ Assign the values in the matrix to the rows with names 8th and 9th. Because these rows don't exist yet, R creates them automatically.

Actually, you don't need to construct the matrix first; you can just use a vector instead. Exactly as with matrices, data frames are filled column-wise. So, the following code gives you exactly the same result:

```
> baskets.df[c("8th", "9th"), ] <- c(3, 8, 9, 4)
```

This process works only for data frames, though. If you try to do the same thing with matrices, you get an error. In the case of matrices, you can only use indices that exist already in the original object.

You have multiple equally valid options for adding observations to a data frame. Which option you choose depends on your personal choice and the situation. If you have a matrix or data frame with extra observations, you can use `rbind()`. If you have a vector with row names and a set of values, using the indices may be easier.

Adding variables to a data frame

A data frame also can be extended with new variables. You may, for example, get data from another player on Granny's team. Or you may want to calculate a new variable from the other variables in the dataset, like the total sum of baskets made in each game (see also Chapter 13).

Adding a single variable

There are three main ways of adding a variable. Similar to the case of adding observations, you can use either the cbind() function or the indices. We illustrate both methods later in this section.

You also can use the dollar sign to add an extra variable. Imagine that Granny asked you to add the number of baskets of her friend Gabrielle to the data frame. First, you would create a vector with that data like this:

```
> baskets.of.Gabrielle <- c(11, 5, 6, 7, 3, 12, 4, 5, 9)
```

To create an extra variable named Gabrielle with that data, you simply do the following:

```
> baskets.df$Gabrielle <- baskets.of.Gabrielle
```

If you want to check whether this worked, but you don't want to display the complete data frame, you could use the head() function. This function takes two arguments: the object you want to display and the number of rows you want to see. To see the first four rows of the new data frame, baskets.df, use the following code:

```
> head(baskets.df, 4)
    Granny Geraldine Gabrielle
1st     12         5        11
2nd      4         4         5
3rd      5         2         6
4th      6         4         7
```

Adding multiple variables using cbind

As we mention earlier, you can pretend your data frame is a matrix and use the cbind() function to do this. Unlike when you use rbind() on data frames, you don't even need to worry about the row or column names. Let's create a new data frame with the goals for Gertrude and Guinevere. To combine both into a data frame, try:

```
> new.df <- data.frame(
+    Gertrude  = c(3, 5, 2, 1, NA, 3, 1, 1, 4),
+    Guinevere = c(6, 9, 7, 3, 3, 6, 2, 10, 6)
+ )
```

Although the row names of the data frames `new.df` and `baskets.df` differ, R will ignore this and just use the row names of the first data frame in the `cbind()` function, as you can see from the output of the following code:

```
> head(cbind(baskets.df, new.df), 4)
    Granny Geraldine Gabrielle Gertrude Guinevere
1st     12         5        11        3         6
2nd      4         4         5        5         9
3rd      5         2         6        2         7
4th      6         4         7        1         3
```

When using a data frame or a matrix with column names, R will use those as the names of the variables. If you use `cbind()` to add a vector to a data frame, R will use the vector's name as a variable name unless you specify one yourself, as you did with `rbind()`.

If you bind a matrix without column names to the data frame, R automatically uses the column numbers as names. That will cause a bit of trouble though, because plain numbers are invalid object names and, hence, more difficult to use as variable names. In this case, you'd better use the indices.

Whenever you want to use a data frame and don't want to continuously have to type its name followed by $, you can use the functions `with()` and `within()`, as explained in Chapter 13. With the `within()` function, you also can easily add variables to a data frame.

Combining Different Objects in a List

In the previous sections, you discover how much data frames and matrices are treated alike by many R functions. But contrary to what you would expect, data frames are not a special case of matrices but a special case of lists. A *list* is a very general and flexible type of object in R. Many statistical functions you use in Chapters 14 and 15 give a list as output. Lists also can be very helpful to group different types of objects, or to carry out operations on a complete set of different objects. You do the latter in Chapter 9.

Creating a list

It shouldn't come as a surprise that you create a list with the `list()` function. You can use the `list()` function in two ways: to create an unnamed list or to create a named list. The difference is small; in both cases, think of a list as a big box filled with a set of bags containing all kinds of different stuff. If these bags are labeled instead of numbered, you have a named list.

Creating an unnamed list

Creating an unnamed list is as easy as using the `list()` function and putting all the objects you want in that list between the `()`. In the previous sections, you worked with the matrix `baskets.team`, containing the number of baskets Granny and Geraldine scored this basketball season. If you want to combine this matrix with a character vector indicating which season we're talking about here, try:

```
> baskets.list <- list(baskets.team, "2010-2011")
```

If you look at the object `baskets.list`, you see the following output:

```
> baskets.list
[[1]]
          1st 2nd 3rd 4th 5th 6th
Granny     12   4   5   6   9   3
Geraldine   5   4   2   4  12   9
[[2]]
[1] "2010-2011"
```

The object `baskets.list` contains two components: the matrix and the season. The numbers between the `[[ ]]` indicate the "bag number" of each component.

Creating a named list

In order to create a labeled, or *named,* list, you simply add the labels before the values between the `()` of the `list()` function, like this:

```
> baskets.nlist <- list(scores = baskets.team, season = "2010-2011")
```

This is exactly the same thing you do with data frames in the "Manipulating Values in a Data Frame" section, earlier in this chapter. And that shouldn't surprise you, because data frames are, in fact, a special kind of named list.

If you look at the named list `baskets.nlist`, you see the following output:

```
> baskets.nlist
$scores
          1st 2nd 3rd 4th 5th 6th
Granny     12   4   5   6   9   3
Geraldine   5   4   2   4  12   9

$season
[1] "2010-2011"
```

Now the `[[ ]]` moved out and made a place for the `$` followed by the name of the component. In fact, this begins to look a bit like a data frame.

Data frames are nothing but a special type of named list, so all the tricks in the following sections can be applied to data frames as well. We repeat: *All the tricks in the following sections — really, all of them — can also be used on data frames.*

Playing with the names of components

Just as with data frames, you access the the names of a list using the `names()` function, like this:

```
> names(baskets.nlist)
[1] "scores" "season"
```

This means that you also can use the `names()` function to add names to the components or change the names of the components in the list in much the same way you do with data frames.

Getting the number of components

Data frames are lists, so it's pretty obvious that the number of components in a list is considered the length of that list. So, to know how many components you have in `baskets.list`, you simply do the following:

```
> length(baskets.list)
[1] 2
```

Extracting components from lists

The display of both the unnamed list `baskets.list` and the named list `baskets.nlist` show already that the way to access components in a list differs from the methods you've used until now.

That's not completely true, though. In the case of a named list, you can access the components using the $, as you do with data frames. For both named and unnamed lists, you can use two other methods to access components in a list:

- ✔ Using [[ ]] gives you the component itself.
- ✔ Using [ ] gives you a list with the selected components.

Using [[ ]]

If you need only a single component and you want the component itself, you can use [[ ]], like this:

```
> baskets.list[[1]]
          1st 2nd 3rd 4th 5th 6th
Granny     12   4   5   6   9   3
Geraldine   5   4   2   4  12   9
```

If you have a named list, you also can use the name of the component as an index, like this:

```
> baskets.nlist[["scores"]]
          1st 2nd 3rd 4th 5th 6th
Granny     12   4   5   6   9   3
Geraldine   5   4   2   4  12   9
```

In each case, you get the component itself returned. Both methods give you the original matrix baskets.team.

You can't use logical vectors or negative numbers as indices when using [[ ]]. You can use only a single value — either a (positive) number or a component name.

Using [ ]

You can use [ ] to extract either a single component or multiple components from a list, but in this case the outcome is always a list. [ ] is more flexible than [[ ]], because you can use all the tricks you also use with vector and matrix indices. [ ] can work with logical vectors and negative indices as well.

So, if you want all components of the list baskets.list except for the first one, you can use the following code:

```
> baskets.list[-1]
[[1]]
[1] "season 2010-2011"
```

Or if you want all components of baskets.nlist where the name contains "season", you can use the following code:

```
> baskets.nlist[grepl("season", names(baskets.nlist))]
$season
[1] "2010-2011"
```

Note that, in both cases, the returned value is a list, even if it contains only one component. R simplifies arrays by default, but the same doesn't count for lists.

Changing the components in lists

Much like all other objects we cover up to this point, lists aren't static objects. You can change components, add components, and remove components from them in a pretty straightforward manner.

Changing the value of components

Assigning a new value to an component in a list is pretty straightforward. You use either the $ or the [[ ]] to access that component, and simply assign a new value. If you want to replace the scores in the list baskets.nlist with the data frame baskets.df, for example, you can use any of the following options:

```
> baskets.nlist[[1]] <- baskets.df
> baskets.nlist[["scores"]] <- baskets.df
> baskets.nlist$scores <- baskets.df
```

If you use [ ], the story is a bit different. You can change components using [ ] as well, but you have to assign a list of components. So, to do the same as the preceding options using [ ], you need to use following code:

```
> baskets.nlist[1] <- list(baskets.df)
```

All these options have exactly the same result, so you may wonder why you would ever use the last option. Simple: Using [ ] allows you to change more than one component at once. You can change both the season and the scores in baskets.list with the following line of code:

```
> baskets.list[1:2] <- list(baskets.df, "2009-2010")
```

This line replaces the first component in baskets.list with the value of baskets.df, and the second component of baskets.list with the character value "2009-2010".

Removing components

Removing components is even simpler: Just assign the NULL value to the component. In most cases, the component is simply removed. To remove the first component from baskets.nlist, you can use any of these (and more) options:

```
> baskets.nlist[[1]] <- NULL
> baskets.nlist$scores <- NULL
> baskets.nlist["scores"] <- NULL
```

Using single brackets, you again have the possibility of deleting more than one component at once. Note that, in this case, you don't have to create a list with the value NULL first. To the contrary, if you were to do so, you would give the component the value NULL instead of removing it, as shown in the following example:

```
> baskets.nlist <- list(scores = baskets.df, season = "2010-2011")
> baskets.nlist["scores"] <- list(NULL)
> baskets.nlist
$scores
NULL

$season
[1] "2010-2011"
```

Adding extra components using indices

In the section "Adding variables to a data frame," earlier in this chapter, you use either the $ or indices to add extra variables. Lists work the same way; to add a component called players to the list baskets.nlist, you can use any of the following options:

```
> baskets.nlist$players <- c("Granny", "Geraldine")
> baskets.nlist[["players"]] <- c("Granny", "Geraldine")
> baskets.nlist["players"] <- list(c("Granny", "Geraldine"))
```

Likewise, to add the same information as a third component to the list baskets.list, you can use any of the following options:

```
> baskets.list[[3]] <- c("Granny", "Geraldine")
> baskets.list[3] <- list(c("Granny", "Geraldine"))
```

These last options require you to know exactly how many components a list has before adding an extra component. If baskets.list contained three components already, you would overwrite that one instead of adding a new one.

Combining lists

If you wanted to add components to a list, it would be nice if you could do so without having to worry about the indices at all. For that, the only thing you need is a function you use extensively in all the previous chapters, the c() function.

That's right, the c() function — which is short for concatenate — does a lot more than just creating vectors from a set of values. The c() function can combine different types of objects and, thus, can be used to combine lists into a new list as well.

In order to be able to add the information about the players, you have to create a list first. To make sure you have the same output, you have to rebuild the original baskets.list as well. You can do both using the following code:

```
> baskets.list <- list(baskets.team, "2010-2011")
> players <- list(rownames(baskets.team))
```

Then you can combine this players list with the list goal.list like this:

```
> c(baskets.list, players)
[[1]]
          1st 2nd 3rd 4th 5th 6th
Granny     12   4   5   6   9   3
Geraldine   5   4   2   4  12   9

[[2]]
[1] "2010-2011"

[[3]]
[1] "Granny"    "Geraldine"
```

If any of the lists contains names, these names are preserved in the new object as well.

Reading the output of str() for lists

Many people who start with R get confused by lists in the beginning. There's really no need for that — a list has only two important parts: the components and the names. And in the case of unnamed lists, you don't even have to worry about the latter. But if you look at the structure of `baskets.list` in the following output, you can see why people often shy away from lists.

```
> str(baskets.list)
List of 2
 $ : num [1:2, 1:6] 12 5 4 4 5 2 6 4 9 12 ...
  ..- attr(*, "dimnames")=List of 2
  .. ..$ : chr [1:2] "Granny" "Geraldine"
  .. ..$ : chr [1:6] "1st" "2nd" "3rd" "4th" ...
 $ : chr "2010-2011"
```

This really looks like some obscure code used by the secret intelligence services during World War II. Still, when you know how to read it, it's pretty easy to read. So let's split up the output to see what's going on here:

✔ The first line simply tells you that `baskets.list` is a list with two components.

✔ The second line contains a $, which indicates the start of the first component. The rest of that line you should be able to read now: It tells you that this first component is a numeric matrix with two rows and six columns (see the previous sections on matrices).

✔ The third line is preceded by `..`, indicating that this line also belongs to the first component. If you look at the output of `str(baskets.team)` you see this line and the following two as well. R keeps the row and column names of a matrix in an attribute called `dimnames`. In the sidebar "Playing with attributes," earlier in this chapter, you manipulate those yourself. For now, you have to remember only that an attribute is an extra bit of information that can be attached to almost any object in R.

The `dimnames` attribute is by itself again a list.

✔ The fourth and fifth lines tell you that this list contains two components: a character vector of length 2 and one of length 6. R uses the `..` only as a placeholder, so you can read from the indentation which lines belong to which component.

✔ Finally, the sixth line starts again with a $ and gives you the structure of the second component — in this case, a character vector with only one value.

If you look at the output of the `str(baskets.nlist)`, you get essentially the same thing. The only difference is that R now puts the name of each component right after the $.

In many cases, looking at the structure of the output from a function can give you a lot of insight into which information is contained in that object. Often, these objects are lists, and the piece of information you're looking for is buried somewhere in that list.

Seeing the forest through the trees

Working with lists in R is not difficult when you're used to it, and lists offer many advantages. You can keep related data neatly together, avoiding an overload of different objects in your workspace. You have access to powerful functions to apply a certain algorithm on a whole set of objects at once. Above all, lists allow you to write flexible and efficient code in R.

Yet, many beginning programmers shy away from lists because they're overwhelmed by the possibilities. R allows you to manipulate lists in many different ways, but often it isn't clear what the best way to perform a certain task is.

Very likely, you'll get into trouble at some point by missing important details, but don't let that scare you. There are a few simple rules that can prevent much of the pain:

- ✔ **If you can name the components in a list, do so.** Working with names always makes life easier, because you don't have to remember the order of the components in the list.

- ✔ **If you need to work on a single component, always use either** [[ ]] **or $.**

- ✔ **If you need to select different components in a list, always use** []. Having named components can definitely help in this case.

- ✔ **If you need a list as a result of a command, always use** [].

- ✔ **If the components in your list have names, use them!**

- ✔ **If in doubt, consult the Help files.**

Part III
Coding in R

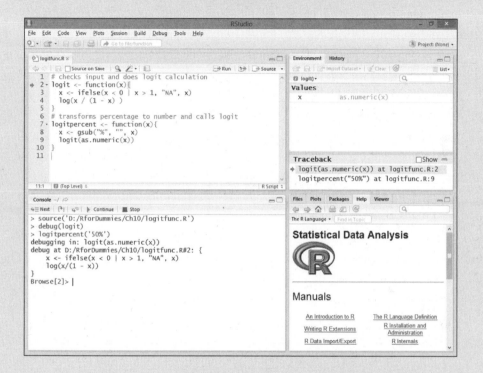

In this part . . .

- ✔ Forming repeating tasks into functions.
- ✔ Creating loops.
- ✔ Squashing bugs.
- ✔ Finding help.
- ✔ Visit www.dummies.com/extras/r for great Dummies content online.

Chapter 8

Putting the Fun in Functions

. . . .

. . . .

*A*utomating your work is probably the number one reason why you use a programming language. You can get pretty far with the built-in functions of R in combination with scripts, but scripts aren't very flexible when dealing with variable input. Luckily, R allows you to write your own custom functions (for example, to automate the cleaning of your data, to apply a series of analyses with one command, or to construct custom plots).

The functions that you write yourself are essentially the same as many of the built-in functions — they are first-class citizens. In this chapter, you discover how to write and work with functions in R.

Moving from Scripts to Functions

Going from a script to a function doesn't take much effort at all. A function is essentially a piece of code that is executed consecutively and without interruption. In that way, a function doesn't differ that much from a script run using the `source()` function, as we explain in Chapter 2.

However, a function has two very nice advantages over scripts:

✔ Functions can work with variable input, so you use it with different data.

✔ Functions return the output as an object, so you can work with the result of that function.

The best way to learn to swim is by jumping in the deep end, so next you write a function to see how easy this is in R.

Making the script

Suppose you want to present fractional numbers (for example, 1/2) as percentages, nicely rounded to one decimal digit. Here's how to achieve that:

1. **Multiply the fractional numbers by 100.**

2. **Round the result to one decimal place.**

   You can use the round() function to do this (see Chapter 4).

3. **Paste a percentage sign after the rounded number.**

   The paste() function is at your service to fulfill this task (see Chapter 5).

4. **Print the result.**

   The print() function does this.

You can easily translate these steps into a little script for R. So, open a new script file in your editor and type the following code:

```
x <- c(0.458, 1.6653, 0.83112)
percent <- round(x * 100, digits = 1)
result <- paste(percent, "%", sep = "")
print(result)
```

If you save this script as a script file — for example, pastePercent.R — you can now call this script in the console (as shown in Chapter 2) with the following command:

```
> source("pastePercent.R")
[1] "45.8%"  "166.5%" "83.1%"
```

That works splendidly, as long as you want to see the same three numbers every time you call the script. But using the script for other data would be mildly inconvenient, because you would have to change the script every time.

In most editors, you also can *source* a script (send a complete script file to the R console) with one simple click. In RStudio, this is done by clicking the Source button or by pressing Ctrl+Shift+S for sourcing without echo, and Ctrl+Shift+Enter for sourcing with echo.

Transforming the script

To make this script into a function, you need to do a few things. Imagine the script as a little factory that takes the raw numeric material and polishes it up to shiny percentages every mathematician will crave.

First, you have to construct the factory building, preferably with an address so people would know where to send their numbers. Then you have to install a front gate so you can get the raw numbers in. Next, you create the production line to transform those numbers. Finally, you have to install a back gate so you can send your shiny percentages into the world.

To build your factory, change the script to the following code:

```
addPercent <- function(x){
  percent <- round(x * 100, digits = 1)
  result <- paste(percent, "%", sep = "")
  return(result)
}
```

Take a closer look at the different parts that make up this little factory. The function has the following elements:

- ✔ The keyword `function` always must be followed by parentheses. It tells R that what comes next is a function.

- ✔ The parentheses after `function` form the front gate, or *argument list,* of your function. Between the parentheses, the arguments to the function are given. In this case, there's only one argument, named x.

- ✔ The braces, { }, can be seen as the walls of your function. Everything between the braces is part of the assembly line, or the *body* of your function.

- ✔ The `return()` statement is the back gate of your function. The object you put between the parentheses is returned from inside the function to your workspace. You can put only one object between the parentheses.

If you put all this together, you get a complete function, but R doesn't know where to find it yet. So, you use the assignment operator `<-` to put this complete function into an object named `addPercent`. This is the address R can send numbers to for transformation. Now the function has a nice name and is ready to use.

You can't specify in the argument list that x should be a numeric vector. For example, if you try to use a character vector as a value for x, the multiplication inside the body will throw an error because you can't multiply characters by a number. If you want to control which type of object is given as an argument, you have to do so manually, in the body of the function. (You see examples of that in Chapters 9 and 10.)

Using the function

Save the script again, and load it into the console using the `source()` function displayed earlier. Now you see . . . nothing. R doesn't let you know by itself that you created a function, but it's there in the global environment, as you can confirm by using `ls()`:

```
> ls()
[1] "addPercent" "percent"      "result"       "x"
```

If you create a function and load it in the global environment by sourcing the script containing the function, this function becomes an object in the global environment and can, thus, be found using `ls()` and — if necessary — removed using `rm()`.

Formatting the numbers

The output of `ls()` tells you the function is there, so you should be able to use it. You can now create the most astonishing percentages by using the `addPercent()` function like this:

```
> new.numbers <- c(0.8223, 0.02487, 1.62, 0.4)
> addPercent(new.numbers)
[1] "82.2%" "2.5%"  "162%"  "40%"
```

Actually, you could use the code `sprintf("%1.1f%%", 100*x)` instead of the `addPercent()` function for a very similar result. C coders will recognize `sprintf()` immediately and agree that it's both incredibly versatile and complex. The function comes with a very long Help page that's definitely worth reading if you need to format values often. If not, save yourself the headache.

Playing with function objects

Because a function in R is just another object, you can manipulate it much the same way as you manipulate other objects. You can assign the function to a new object and effectively copy it like this:

```
> ppaste <- addPercent
```

Now `ppaste` is a function as well that does exactly the same as `addPercent`. Note that you don't add parentheses after `addPercent` in this case.

If you add the parentheses, you call the function and put the result of that call in `ppaste`. If you don't add the parentheses, you refer to the function object itself without calling it. This difference is important when you use functions as arguments (see the "Using functions as arguments" section, later in this chapter).

You can print the content of a function by simply typing its name at the prompt, like this:

```
> ppaste
function(x){
  percent <- round(x * 100, digits = 1)
  result <- paste(percent, "%", sep = "")
  return(result)
}
```

So, the assignment to `ppaste` actually copied the function code of `addPercent` into a new object.

That's all cool, but it also means that you can effectively erase a function if you accidentally use the same name for another object. Or you could lose data if you accidentally gave the same name as your data object to a function. There's no undo button in R, so pay attention to the names you choose.

Luckily, this problem doesn't occur with the base R functions and functions contained in packages. Although it's not a good idea, you could, for example, name a vector `sum` and still be able to use the `sum()` function afterward. When you use `sum()` as a function, R only searches for functions with that name and disregards all other objects with the same name.

Reducing the number of lines

Not all elements mentioned in the "Transforming the script" section, earlier in this chapter, are required. In fact, the `return()` statement is optional, because, by default, R always returns the value of the last line of code in the function body.

Returning values by default

Suppose you forgot to add `return(result)` in the `addPercent()` function. What would happen then? You can find out if you delete the last line of the `addPercent()` function, save the file, and source it again to load it into the workspace.

Any change you make to a function will take effect only after you send the adapted code to the console. This will effectively overwrite the old function object by a new one.

If you try `addPercent(new.numbers)` again, you see . . . nothing. Apparently, the function doesn't do anything anymore — but this is an illusion, as you can see with the following code:

```
> print( addPercent(new.numbers) )
 [1] "82.2%"  "2.5%"   "162%"   "40%"
```

In this case, the last line of the function returns the value of `result` *invisibly*, which is why you see it only if you specifically ask to print it. The value is returned invisibly due to the assignment in the last line. Because this isn't really practical, you can drop the assignment in the last line and change the function code to the following:

```
addPercent <- function(x){
  percent <- round(x * 100, digits = 1)
  paste(percent, "%", sep = "")
}
```

This function works again as before. It may look like `return()` is utterly useless, but you really need it if you want to exit the function before the end of the code in the body. For example, you could add a line to the `addPercent` function that checks whether x is numeric, and if not, returns `NULL`, like this:

```
addPercent <- function(x){
  if( !is.numeric(x) ) return(NULL)
  percent <- round(x * 100, digits = 1)
  paste(percent, "%", sep = "")
}
```

In Chapter 9, we explain how to use `if()` conditions. In Chapter 10, you meet the functions you need to throw your own warnings and errors.

Breaking the walls

The braces, `{ }`, form the proverbial wall around the function, but in some cases you can drop them as well. Suppose you want to calculate the odds from a proportion. The odds of something happening is no more than the chance it happens divided by the chance it doesn't happen. So, to calculate the odds, you can write a function like this:

```
> odds <- function(x) x / (1-x)
```

Even without the braces or `return()` statement, this works perfectly fine, as you can see in the following example:

```
> odds(0.8)
[1] 4
```

If a function consists of only one line of code, you can just add that line after the argument list without enclosing it in braces. R will see the code after the argument list as the body of the function.

You could do the same with the `addPercent()` function by nesting everything like this:

```
> addPercent <- function(x) paste(round(x * 100, digits = 1), "%", sep = "")
```

That's a cunning plan to give the next person reading that code a major headache. It's a bit less of a cunning plan if that next person is you, though, and chances are, it will be.

Saving space in a function body is far less important than keeping the code readable, because saving space gains you nothing. Constructs like the odds() function are useful only in very specific cases. You find examples of this in the "Using anonymous functions" section, later in this chapter, as well as in Chapter 13. But for now, remember that using braces { } is a good practice, even when you have only a single line of code in the body of your function.

Using Arguments the Smart Way

In Chapter 3, you saw how to specify arguments in a function call. To summarize:

- Arguments are always *named* when you define the function. But when you call the function, you don't have to specify the name of the argument if you give them in the order in which they appear in the argument list of a function.
- Arguments can be *optional,* in which case you don't have to specify a value for them.
- Arguments can have a *default* value, which is used if you didn't specify a value for that argument yourself.

Not only can you use as many arguments as you like — or as is feasible, at least — but you can very easily pass arguments on to functions inside the body of your own function with the simply genius *dots* argument. Fasten your seat belts — we're off to make some sweet R magic.

Adding more arguments

The argument list of the addPercent() function doesn't really look much like a list yet. Actually, the only thing you can do for now is tell the function which number you want to see converted. It serves perfectly well for this little function, but you can do a lot more with arguments than this.

The addPercent() function automatically multiplies the numbers by 100. This is fine if you want to convert fractions to percentages, but if the calculated numbers are percentages already, you would have to divide these numbers first by 100 to get the correct result, like this:

```
> percentages <- c(58.23, 120.4, 33)
> addPercent(percentages / 100)
[1] "58.2%"  "120.4%"  "33%"
```

That's quite a way around, but you can avoid this by adding another argument to the function that controls the multiplication factor.

Adding the mult argument

You add extra arguments by including them between the parentheses after the function keyword. All arguments are separated by commas. To add an argument mult that controls the multiplication factor in your code, you change the function like this:

```
addPercent <- function(x, mult){
  percent <- round(x * mult, digits = 1)
  paste(percent, "%", sep = "")
}
```

Now you can specify the mult argument in the call to addPercent(). If you want to use the percentages vector from the previous section, you use the addPercent() function, like this:

```
> addPercent(percentages, mult = 1)
[1] "58.2%"  "120.4%"  "33%"
```

Adding a default value

Adding an extra argument gives you more control over what the function does, but it also introduces a new problem. If you don't specify the mult argument in the addPercent() function, you get the following result:

```
> addPercent(new.numbers)
Error in x * mult : 'mult' is missing
```

Because you didn't specify the mult argument, R has no way of knowing which number you want to multiply x by, so it stops and tells you it needs more information. That's pretty annoying, because it also means you would have to specify mult=100 every time you used the function with fractions. Specifying a default value for the argument mult takes care of this.

You specify default values for any argument in the argument list by adding the = sign and the default value after the respective argument.

To get the wanted default behavior, you adapt `addPercent()` like this:

```
addPercent <- function(x, mult = 100){
  percent <- round(x * mult, digits = 1)
  paste(percent, "%", sep = "")
}
```

Now the argument works exactly the same as arguments with a default value from base R functions. If you don't specify the argument, the default value of 100 is used. If you do specify a value for that argument, that value is used instead. So, in the case of `addPercent()`, you can now use it as shown in the following example:

```
> addPercent(new.numbers)
[1] "82.2%" "2.5%"  "162%"  "40%"
> addPercent(percentages, 1)
[1] "58.2%"  "120.4%" "33%"
```

You don't have to specify the names of the arguments if you give them in the same order as they're given in the argument list. This works for all functions in R, including those you create yourself.

Conjuring tricks with dots

The `addPercent()` function rounds every percentage to one decimal place. To add another argument to specify the number of digits used in rounding, you can specify the argument explicitly as you did for the `mult` argument in the previous section. However, if you have many arguments to pass on to other functions inside the body, you'll end up with quite a long list of arguments.

R has a genius solution for this: the dots (`. . .`) argument. You can see the dots argument as an extra gate in your little function. Through that gate, you drop additional resources (arguments) immediately at the right spot in the production line (the body) without the hassle of having to check everything at the main gate.

You normally use the dots argument by adding it at the end of the argument list of your own function and at the end of the arguments for the function you want to pass arguments to.

To pass any argument to the round() function inside the body of addPercent, adapt the code of the latter as follows:

```
addPercent <- function(x, mult = 100, ...){
  percent <- round(x * mult, ...)
  paste(percent, "%", sep = "")
}
```

Now you can specify the digits argument for round() in the addPercent() call like this:

```
> addPercent(new.numbers, digits = 2)
[1] "82.23%" "2.49%"  "162%"   "40%"
```

You don't have to specify any argument if the function you pass the arguments to doesn't require it. You can use addPercent() as before:

```
> addPercent(new.numbers)
[1] "82%"  "2%"   "162%" "40%"
```

Notice that the outcome isn't the same as it used to be. The numbers are rounded to integers and not to the first decimal.

If you don't specify an argument in lieu of the dots, the function where the arguments are passed to uses its own default values. If you want to specify different default values, you'll have to add a specific argument to the argument list instead of using the dots.

So, to get addPercent() to use a default rounding to one decimal, you have to use the following code:

```
addPercent <- function(x, mult = 100, digits = 1){
  percent <- round(x * mult, digits = digits)
  paste(percent, "%", sep = "")
}
```

You don't have to give the argument in the argument list the same name as the argument used by round(). You can use whatever name you want, as long as you place it in the right position within the body. However, if you can use names for arguments that also are used by native functions within R, it'll be easier for people to understand what the argument does without having to look at the source code.

R won't complain if you use the dots argument in more than one function within the body, but before passing arguments to more than one function in the body, you have to be sure this won't cause any trouble. R passes *all* extra arguments to *every* function, and — if you're lucky — complains about the resulting mess afterward.

Using functions as arguments

You read that correctly. In R, you can pass a function itself as an argument. In the "Playing with function objects" section, earlier in this chapter, you saw that you can easily assign the complete code of a function to a new object. In much the same way, you also can assign the function code to an argument. This opens up a complete new world of possibilities. We show you only a small piece of that world in this section.

Applying different ways of rounding

In Chapter 4, we show you different options for rounding numbers. The `addPercent()` function uses `round()` for that, but you may want to use one of the other options — for example, `signif()`. The `signif()` function doesn't round to a specific number of decimals; instead, it rounds to a specific number of digits (see Chapter 4). You can't use it before you call `addPercent()`, because the `round()` function in that body will mess everything up again.

Of course, you could write a second function specifically for that, but there's no need to do so. Instead, you can just adapt `addPercent()` in such a way that you simply give the function you want to use as an argument:

```
addPercent <- function(x, mult = 100, FUN = round, ...){
  percent <- FUN(x * mult, ...)
  paste(percent, "%", sep = "")
}
```

This really couldn't be easier: You add an argument to the list — in this case, `FUN` — and then you can use the name of that argument as a function. Also, specifying a default value works exactly the same as with other arguments; just specify the default value — in this case, `round` — after an = sign.

If you want to use `signif()` now for rounding the numbers to three digits, you can easily do that using the following call to `addPercent()`:

```
> addPercent(new.numbers, FUN = signif, digits = 3)
[1] "82.2%" "2.49%" "162%"  "40%"
```

What happens here?

1. As before, R takes the vector `new.numbers` and multiplies it by 100, because that's the default value for `mult`.

2. R assigns the function code of `signif` to FUN, so now `FUN()` is a perfect copy of `signif()` and works exactly the same way.

3. R takes the argument `digits` and passes it on to `FUN()`.

Note the absence of parentheses in the argument assignment. If you added the parentheses there, you would assign the result of a call to `signif()` instead of the function itself. R would interpret `signif()`, in that case, as a nested function, and that's not what you want. Plus, R would throw an error because, in that case, you call `signif()` without arguments, and R doesn't like that.

Using anonymous functions

You can, of course, use any function you want for the FUN argument. In fact, that function doesn't even need to have a name, because you effectively copy the code. So, instead of giving a function name, you can just add the code as an argument as a nameless or *anonymous* function. An anonymous function is a function without a name.

Suppose you have the quarterly profits of your company in a vector like this:

```
> profits <- c(2100, 1430, 3580, 5230)
```

Your manager asks you to report the profit for each quarter relative to the total for the year, and, of course, you want to use your new `addPercent()` function. To calculate the relative profits in percent, you could write a `rel.profit()` function like this:

```
> rel.profit <- function(x) round(x / sum(x) * 100)
```

But you don't have to. Instead, you can just use the function body itself as an argument, as in the following example:

```
> addPercent(profits,
+            FUN = function(x) round(x / sum(x) * 100) )
[1] "17%" "12%" "29%" "42%"
```

Of course, this isn't the optimal way of doing this specific task. You could easily have gotten the same result with the following code:

```
> addPercent(profits / sum(profits))
[1] "17%" "12%" "29%" "42%"
```

In some cases, this construct with anonymous functions is really a treat, especially when you want to use functions that can be written in only a little code and aren't used anywhere else in your script. (You find more — and better — examples of anonymous functions in Chapter 13.)

Matching functions

In the examples in the "Using functions as arguments" section, you effectively pass the code of a function as an argument. This also means that if you have an object with the same name as the function you want to use, this whole construct won't work. Suppose you had the not-so-smart idea of creating a vector with the relative gain of a couple rounds of poker like this:

```
> round <- c(0.48, -0.52, 1.88)
```

If you tried to call `addPercent()` with the FUN argument on this vector, you'd get the following error:

```
> addPercent(round, FUN = round)
Error in addPercent(round, FUN = round) :
  could not find function "FUN"
```

Instead of passing the code of the `round` function, R passes the vector `round` as the FUN argument. To avoid these kinds of problems, you can use a special function, `match.fun()`, in the body of `addPercent()`, like this:

```
addPercent <- function(x, mult = 100,
        FUN, ...){
  FUN <- match.fun(FUN)
  percent <- FUN(x * mult, ...)
  paste(percent, "%", sep = "")
}
```

This function looks for a function that matches the name `round` and copies that code into the FUN argument instead of the vector `round`. As an added bonus, `match.fun()` also allows you to use a character object as the argument, so specifying `FUN = 'round'` now works as well. All native R functions use `match.fun()` for this purpose, and we can only advise you to do the same if you write code that will be used by other people. But passing functions works fine without using `match.fun()`, as long as you use sensible names for the other objects in your global environment.

Finally, clean up the round object so you don't get into trouble later on:

```
> rm(round)
```

Coping with Scoping

In the previous chapters, you work solely in the workspace. Every object you create ends up in this environment, which is called the *global environment*. The global environment is the universe of the R user where everything happens.

R gurus will tell you that this "universe" is actually contained in another "universe" and that one in yet another, and so on — but that "outer space" is a hostile environment suited only to daring coders without fear of breaking things. So, there's no need to go there now.

Crossing the borders

In the functions in the previous sections, you work with some objects that you didn't first create in the global environment. You use the arguments x, mult, and FUN as if they're objects, and you create an object percent within the function that you can't find back in the global environment after using the function. So, what's going on?

Creating a test case

Let's find out through a small example. First, create an object x and a small test() function like this:

```
x <- 1:5
test <- function(x){
  cat("This is x:", x, "\n")
  rm(x)
  cat("This is x after removing it:", x, "\n")
}
```

The test() function doesn't do much. It takes an argument x, prints it to the console, removes it, and tries to print it again. You may think this function will fail, because x disappears after the line rm(x). But no, if you try this function it works just fine, as shown in the following example:

```
> test(5:1)
This is x: 5 4 3 2 1
This is x after removing it: 1 2 3 4 5
```

Even after removing x, R still can find another x that it can print. If you look a bit more closely, you see that the x printed in the second line is actually not the one you gave as an argument, but the x you created before in the global environment. How come?

Searching the path

If you use a function, the function first creates a temporary *local environment*. This local environment is *nested* within the global environment, which means that, from that local environment, you also can access any object from the global environment. As soon as the function ends, the local environment is destroyed together with all objects in it.

To be completely correct, a function always creates an environment within the environment it's called from, called the *parent environment*. If you call a function from the global environment, either through a script or by using the command line, this parent environment happens to be the global environment.

You can see a schematic illustration of how the test() function works in Figure 8-1. The big rectangle represents the global environment, and the small rectangle represents the local environment of the test function. In the global environment, you assign the value 1:5 to the object x. In the function call, however, you assign the value 5:1 to the argument x. This argument becomes an object x in the local environment.

Figure 8-1:
How R looks through global and local environments.

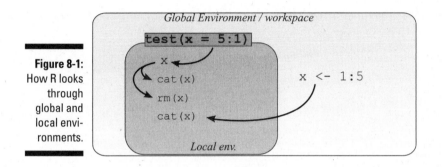

If R sees any object name — in this case, x — mentioned in any code in the function, it first searches the local environment. Because it finds an object x there, it uses that one for the first cat() statement. In the next line, R removes that object x. So, when R reaches the third line, it can't find an object x in the local environment anymore. No problem. R moves up the stack of environments and checks to see if it finds anything looking like an x in the global environment. Because it can find an x there, it uses that one in the second cat() statement.

If you use rm() inside a function, rm() will, by default, delete only objects within the local environment of that function. This way, you can avoid running out of memory when you write functions that have to work on huge datasets. You can immediately remove big temporary objects instead of waiting for the function to do so at the end.

Dispatching to a Method

We want to cover one more thing about functions, because you need it to understand how one function can give a different result based on the type of value you give the arguments. R has a genius system, called the *generic function system,* allowing you to call different functions using the same name. If you think this is weird, think again about data frames and lists (see Chapter 7). If you print a list in the console, you get the output arranged in rows. On the other hand, a data frame is printed to the console arranged in columns. So, the `print()` function treats lists and data frames differently, but both times you used the same function. Or did you really?

Finding the methods behind the function

It's easy to find out if you used the same function both times — you can just peek inside the function code of `print()` by typing its name at the command line, like this:

```
> print
function (x, ...)
UseMethod("print")
<bytecode: 0x0464f9e4>
<environment: namespace:base>
```

You can safely ignore the two last lines, because they refer to complicated stuff in the "outer space" of R and are used only by R developers. But take a look at the function body — it's only one line!

Functions that don't do much other than passing on objects to the right function are called *generic functions*. In this example, `print()` is a generic function. The functions that do the actual work are called *methods*. So, every method is a function, but not every function is a method.

Using methods with UseMethod

How on earth can that one line of code in the `print()` function do so many complex things like printing vectors, data frames, and lists all in a different way? The answer is contained in the `UseMethod()` function, which is the central function in the generic function system of R. All `UseMethod()` does is tell R to move along and look for a function that can deal with the type of object that is given as the argument `x`.

R does that by looking through the complete set of functions in search of another function that starts with `print` followed by a dot and then the name of the object type.

You can do that yourself by using the command `methods("print")`. The function `methods()` lists all known methods for the generic function of interest. Don't be surprised when you get over 40 different `print()` functions for all kinds of objects. If you want to find all methods for a specific type of object, you can specify the argument `class`. For example, `methods(class = "data.frame")` will give you all methods specific to the class `data.frame`.

Suppose you have a data frame you want to print. R looks up the function `print.data.frame()` and uses that function to print the object you passed as an argument. You also can call that function yourself like this:

```
> small.one <- data.frame(a = 1:2, b = 2:1)
> print.data.frame(small.one)
  a b
1 1 2
2 2 1
```

The effect of that function differs in no way from what you would get if you used the generic `print(small.one)` function instead. That's because `print()` will give the `small.one` to the `print.data.frame()` function to take care of it.

Using default methods

In the case of a list, you may be tempted to look for a `print.list()` function. But it won't work, because the `print.list()` function doesn't exist. This isn't a problem — R ignores the type of the object in that case and looks for a *default method,* `print.default()`.

For many generic functions, there is a default method that's used if no specific method can be found. If there is one, you can recognize the default method by the word *default* after the dot in the function name.

So, if you want to print the data frame as a list, you can use the default method like this:

```
> print.default(small.one)
$a
[1] 1 2
$b
[1] 2 1
attr(,"class")
[1] "data.frame"
```

Object-oriented programming (OOP) in R

The system of method dispatching described in this chapter is called S3, and it's only one of the ways you can work object oriented in R. If you've programmed in an object-oriented language like Java or C++ before, be sure to check out the options for OOP in R. The method dispatching here is only the very start; the S4 methods and reference classes are far more powerful and include many of the necessary principles of OOP, including constructors, inheritance, and the like. But that's a subject so complex that it could fill a whole book.

OOP in R is covered in quite some detail in *Advanced R,* by Hadley Wickham (Chapman and Hall/CRC).

Doing it yourself

All that method dispatching sounds nice, but it may seem mostly like internal affairs of R and not that interesting to know as a user . . . unless you could use that system yourself, of course — and you can!

Adapting the addPercent function

Suppose you want to be able to paste the percent sign to character vectors with the addPercent function. As the function is written in the previous sections, you can't. A character vector will give an error the moment you try to multiply it, so you need another function for that, like the following:

```
addPercent.character <- function(x){
  paste(x, "%", sep = "")
}
```

Note that the type of the object is not vector but character. In the same way, you also have to rename the original addPercent function to addPercent.numeric in your script.

If you use the system of method dispatching, you can keep all functions in one script file if they aren't too big. That way, you have to source only one script file in order to have the whole generic system working.

All you need now is a generic addPercent() function like this:

```
addPercent <- function(x,...){
  UseMethod("addPercent")
}
```

You use only two arguments here: x and the dots (. . .). The use of the dots argument assures that you can still use all the arguments from the addPercent.numeric() function in your call to addPercent(). The extra arguments are simply passed on to the appropriate method via the dots argument, as explained in the "Using Arguments the Smart Way" section, earlier in this chapter.

After sending the complete script file to the console, you can send both character vectors and numeric vectors to addPercent(), like this:

```
> addPercent(new.numbers, FUN = floor)
[1] "82%"  "2%"   "162%" "40%"
> addPercent(letters[1:6])
[1] "a%" "b%" "c%" "d%" "e%" "f%"
```

A detail that might prevent some headaches is that the argument names in methods should match the argument names in the generic. You don't want to have the argument called x in the generic but object in a method.

Adding a default function

If you try to use the small data frame you made in the previous section, you will get the following error:

```
> addPercent(small.one)
Error in UseMethod("addPercent") :
  no applicable method for 'addPercent' applied to an object of class "data.
           frame"
```

That's a rather complicated way of telling you that there's no method for a data frame. There's no need for a data frame method either, but it may be nice to point out to the users of your code that they should use a vector instead. So, you can easily add a default method exactly like R does:

```
addPercent.default <- function(x){
   message('You should try a numeric or character vector.')
}
> addPercent(small.one)
```

This default method doesn't do much apart from printing a message, but at least that message is a bit easier to understand than the one R spits out. Sometimes it's apparent that the error messages of R aren't always written with a normal end-user in mind!

Chapter 9

Controlling the Logical Flow

- -

In This Chapter

▶ Making choices based on conditions

▶ Looping over different values

▶ Applying functions row-wise and column-wise

▶ Applying functions over values, variables, and list components

- -

A function can be nothing more than a simple sequence of actions, but these kinds of functions are highly inflexible. Often, you want to make choices and take action dependent on a certain value.

Choices aren't the only things that can be useful in functions. Loops can prevent you from having to rewrite the same code over and over again. If you want to calculate the summary statistics on different datasets, for example, you can write the code to calculate those statistics and then tell R to carry out that code for all the datasets you had in mind.

R has a very specific mechanism for looping over values that combines a lot of power with a minimum of unwanted side effects. Instead of using a classic loop structure, you use a function to *apply* another function on any of the objects discussed in the previous chapters. This way of looping over values is one of the features that distinguish R from many other programming languages.

In this chapter, we cover the R tools to create loops and make decisions.

Note: If a piece of code is not preceded by a prompt (>), it represents an example function that you can copy to your editor and then send to the console (as explained in Chapter 2). All code you normally type directly at the command line is preceded by a prompt.

Making Choices with if Statements

Defining a choice in your code is pretty simple: If a condition is true, then carry out a certain task. Many programming languages let you do that with exactly those words: *if . . . then.* R makes it even easier: You can drop the word *then* and specify your choice in an `if` statement.

An `if` statement in R consists of three elements:

- ✔ The keyword `if`
- ✔ A single logical value between parentheses (or an expression that leads to a single logical value)
- ✔ A block of code between braces that has to be executed when the logical value is `TRUE`

To see how easy this is, look at a very small function, `priceCalculator()`, that calculates the price you charge to a customer based on the hours of work you did for that customer. The function takes the number of hours (hours) and the price per hour (pph) as input. The `priceCalculator()` function could look like this:

```
priceCalculator <- function(hours, pph = 40){
    net.price <- hours * pph
    round(net.price)
}
```

Here's what this code does:

- ✔ With the `function` keyword, you define the function.
- ✔ Everything between the braces is the body of the function (see Chapter 8).
- ✔ Between the parentheses, you specify the arguments hours (without a default value) and pph (with a default value of $40 per hour).
- ✔ You calculate the net price by multiplying hours by pph.
- ✔ The outcome of the last statement in the body of your function is the returned value. In this case, this is the total price rounded to the dollar.

You could drop the argument pph and just multiply hours by 40. But that would mean that if, for example, your colleague uses a different hourly rate, he would have to change the value in the body of the function in order to be able to use it. It's good coding practice to use arguments with default values for any value that can change. Doing so makes a function more flexible and usable.

Now imagine you have some big clients that give you a lot of work. To keep them happy, you decide to give them a reduction of 10 percent on the price per hour for orders that involve more than 100 hours of work. So, if the number of hours worked is larger than 100, you calculate the new price by multiplying the price by 0.9. You can write that almost literally in your code like this:

```
priceCalculator <- function(hours, pph = 40){
    net.price <- hours * pph
    if(hours > 100) {
      net.price <- net.price * 0.9
    }
    round(net.price)
}
```

Copy this code in a script file, and send it to the console to make it available for use. If you try out this function, you can see that the reduction is given only when the number of hours is larger than 100:

```
> priceCalculator(hours = 55)
[1] 2200
> priceCalculator(hours = 110)
[1] 3960
```

An `if` statement in R consists of three elements: the keyword `if`, a single logical value between parentheses, and a block of code between braces that has to be executed when the logical value is TRUE. If you look at the `if` statement in the previous function, you find these three elements. Between the parentheses, you find an expression (`hours > 100`) that evaluates to a single logical value. Between the braces stands one line of code that reduces the net price by 10 percent when the line is carried out.

This construct is the most general way you can specify an `if` statement. But if you have only one short line of code in the code block, you don't have to put braces around it. You can change the complete `if` statement in the function with the following line:

```
if(hours > 100) net.price <- net.price * 0.9
```

The usual way of getting help on a function named, for example, `fun.name` (`?fun.name`) does not work for `if`. To access the built-in help for `if`, you have to quote the function name with *backticks,* the symbol that looks like a backward slanting quote (`'`). This prevents R from reading the `if` as part of a condition statement.

```
?'if'
```

The help function also allows you to pass a character vector with the name of the function. So each of the following statements takes you to the Help page for `if`:

```
?'if'
?"if"
```

Doing Something Else with an if...else Statement

In some cases, you need your function to do something if a condition is true and something else if it is not. You could do this with two `if` statements, but there's an easier way in R: an `if...else` statement. An `if...else` statement contains the same elements as an `if` statement (see the preceding section), and then some extra:

✔ The keyword `else`, placed after the first code block

✔ A second block of code, contained within braces, that has to be carried out if and only if the result of the condition in the `if()` statement is `FALSE`

In some countries, the amount of value added tax (VAT) that has to be paid on certain services depends on whether the client is a public or private organization. Imagine that public organizations pay only 6 percent VAT and private organizations pay 12 percent VAT. You can add an extra argument `public` to the `priceCalculator()` function and adopt it as follows to add the correct amount of VAT:

```
priceCalculator <- function(hours, pph = 40, public = TRUE){
    net.price <- hours * pph
    if(hours > 100) net.price <- net.price * 0.9
    if(public) {
       tot.price <- net.price * 1.06
    } else {
       tot.price <- net.price * 1.12
    }
    round(tot.price)
}
```

If you send this code to the console, you can test the function. For example, if you worked for 25 hours, the following code gives you the different amounts you charge for public and private organizations, respectively:

```
> priceCalculator(25, public = TRUE)
[1] 1060
> priceCalculator(25, public = FALSE)
[1] 1120
```

This works well, but how does it work?

If you look at the `if...else` statement in the previous function, you find these elements. If the value of the argument `public` is TRUE, the total price is calculated as 1.06 times the net price. Otherwise, the total price is 1.12 times the net price.

The `if` statement needs a logical value between the parentheses. Any expression you put between the parentheses is evaluated before being passed on to the `if` statement. So, if you work with a logical value directly, you don't have to specify an expression at all. Using, for example, `if(public == TRUE)` is about as redundant as asking if white snow is white. It would work, but it's bad coding practice.

Also, in the case of an `if...else` statement, you can drop the braces if both code blocks exist of only a single line of code. So, you could just forget about the braces and squeeze the whole `if...else` statement on a single line. Or you could even write it like this:

```
if(public) tot.price <- net.price * 1.06 else
          tot.price <- net.price * 1.12
```

Putting the `else` statement at the end of a line and not the beginning of the next one is a good idea. In general, R reads multiple lines as a single line as long as it's absolutely clear that the command isn't finished yet (see Chapter 3). If you put `else` at the beginning of the second line, R considers the first line finished and complains. You can put `else` at the beginning of a next line only if you do so *within a function* and you source the complete file at once to R.

But you can still make this shorter. The `if` statement works like a function and, hence, it also returns a value. As a result, you can assign that value to an object or use it in calculations. So, instead of recalculating `net.price` and assigning the result to `tot.price` within the code blocks, you can use the `if...else` statement like this:

```
tot.price <- net.price * if(public) 1.06 else 1.12
```

R first evaluates the `if...else` statement, and multiplies the outcome by `net.price`. The result is then assigned to `tot.price`. This differs not one iota from the result of the five lines of code we used for the original `if...else` statement. R allows programmers to be incredibly lazy, er, economical here.

Vectorizing Choices

As we discuss in Chapter 4, vectorization is one of the defining attributes of the R language. R wouldn't be R if it didn't have some kind of vectorized version of an `if...else` statement. If you wonder why on earth you would need such a thing, take a look at the problem discussed in this section.

Looking at the problem

The `priceCalculator()` function still isn't very economical to use. If you have 100 clients, you have to calculate the price for every client separately. Check for yourself what happens if you add, for example, three different amounts of hours as an argument:

```
> priceCalculator(c(25, 110))
[1] 1060 4664
Warning message:
In if (hours > 100) net.price <- net.price * 0.9 :
  the condition has length > 1 and only the first element will be used
```

Not only does R warn you that something fishy is going on, but the result you get is plain wrong. Instead of $4,664, the second client should be charged only $4,198:

```
> priceCalculator(110)
[1] 4198
```

What happened? The warning message should give you a fair idea about what went wrong. An `if` statement can deal only with a single value, but the expression `hours > 100` returns two values, as shown by the following code:

```
>  c(25, 110) > 100
[1] FALSE  TRUE
```

Choosing based on a logical vector

The solution you're looking for is the `ifelse()` function, which is a vectorized way of choosing values from two vectors. This remarkable function takes three arguments:

- ✔ A test vector with logical values
- ✔ A vector with values that should be returned if the corresponding value in the test vector is `TRUE`
- ✔ A vector with values that should be returned if the corresponding value in the test vector is `FALSE`

Understanding how it works

Take a look at the following trivial example:

```
> ifelse(c(1, 3) < 2.5 , 1:2 , 3:4)
[1] 1 4
```

To understand how it works, run over the steps the function takes:

1. **The conditional expression** c(1, 3) < 2.5 **is evaluated to a logical vector.**

2. **The first value of this vector is** TRUE, **because 1 is smaller than 2.5. So, the first value of the result is the first value of the second argument, which is 1.**

3. **The next value is** FALSE, **because 3 is larger than 2.5. Hence,** ifelse() **takes the second value of the third argument (which is 4) as the second value of the result.**

4. **A vector with the selected values is returned as the result.**

Trying it out

To see how this works in the example of the priceCalculator() function, try the function out at the command line in the console. Say you have two clients and you worked 25 and 110 hours for them, respectively. You can calculate the net price with the following code:

```
> my.hours <- c(25, 110)
> my.hours * 40 * ifelse(my.hours > 100, 0.9, 1)
[1] 1000 3960
```

Didn't you just read that the second and third arguments should be a vector? Yes, but the ifelse() function recycles its arguments. And that's exactly what it does here. In the preceding ifelse() function call, you translate the logical vector created by the expression my.hours > 100 into a vector containing the numbers 0.9 and 1 in lieu of TRUE and FALSE, respectively.

Adapting the function

Of course, you need to adapt the priceCalculator() function in such a way that you also can input a vector with values for the argument public. Otherwise, you wouldn't be able to calculate the prices for a mixture of public and private clients. The final function looks like this:

```
priceCalculator <- function(hours, pph = 40, public){
    net.price <- hours * pph
    net.price <- net.price * ifelse(hours > 100 , 0.9, 1)
    tot.price <- net.price * ifelse(public, 1.06, 1.12)
    round(price)
}
```

Next, create a little data frame to test the function. For example:

```
> clients <- data.frame(
+   hours = c(25, 110, 125, 40),
+   public = c(TRUE, TRUE, FALSE, FALSE)
+ )
```

You can now use this data frame as arguments for the `priceCalculator()` function, like this:

```
> with(clients, priceCalculator(hours, public = public))
[1] 1060 4198 5040 1792
```

There you go. Problem solved!

Making Multiple Choices

The `if` and `if...else` statements in the previous section leave you with exactly two options, but life is seldom as simple as that.

Imagine you have some clients abroad. Let's assume that any client abroad doesn't need to pay VAT. This leaves you with three different VAT rates: 12 percent for private clients, 6 percent for public clients, and none for foreign clients.

Chaining if...else statements

The most intuitive way to solve this problem is just to chain the choices. If a client is living abroad, don't charge any VAT. Otherwise, check whether the client is public or private and apply the relevant VAT rate.

If you define an argument `client` for your function that can take the values `"abroad"`, `"public"`, and `"private"`, you could code the previous algorithm like this:

```
if(client == "private"){
  tot.price <- net.price * 1.12      # 12% VAT
} else {
  if(client == "public"){
    tot.price <- net.price * 1.06    # 6% VAT
  } else {
    tot.price <- net.price * 1       # 0% VAT
  }
}
```

With this code, you nest the second `if...else` statement in the first `if...else` statement. That's perfectly acceptable and it will work, but imagine what you would have to do if you had four or even more possibilities. Nesting a statement in a statement in a statement in a statement quickly creates one huge curly mess.

Luckily, R allows you to write all that code a bit more clearly. You can chain the `if...else` statements as follows:

```
if(client == "private"){
    tot.price <- net.price * 1.12
} else if(client == "public"){
    tot.price <- net.price * 1.06
} else {
    tot.price <- net.price
}
```

In this example, the chaining makes a difference of only two braces, but when you have more possibilities, it really makes the difference between readable code and sleepless nights. Note, also, that you don't have to test whether the argument `client` is equal to `"abroad"` (although it wouldn't be wrong to do that). You just assume that if `client` doesn't have any of the two other values, it has to be `"abroad"`.

TIP

In this specific case, you can further simplify the code if you keep in mind that `if` also can return a value, as we explained in the earlier section "Doing Something Else with an if. . .else Statement." Using the same trick as before, you can simplify the preceding code to:

```
tot.price <- net.price *
    if(client == "private") 1.12 else
    if(client == "public") 1.06 else 1
```

Chained `if...else` statements work on a single value at a time. You can't use these chained `if...else` statements in a vectorized way. For that, you can nest multiple `ifelse` statements, like this:

```
VAT <- ifelse(client == "private", 1.12,
        ifelse(client == "public", 1.06, 1)
      )
tot.price <- net.price * VAT
```

This style of code can be very confusing if you have more than three choices, though. The solution to this is to `switch`.

Switching between possibilities

The nested `if...else` statement is especially useful if you have complete code blocks that have to be carried out when a condition is met. But if you need to select values based only on a condition, you have a better option: Use the `switch()` function.

Making choices with switch

In the previous example, you wanted to adjust the VAT rate depending on whether the client is public, private, or lives abroad. You have a list of three possible choices, and for each choice you have a specific VAT rate. You can use the `switch()` function like this:

```
VAT <- switch(client, private = 1.12, public = 1.06, abroad = 1)
```

You construct a `switch()` call as follows:

1. **Give a *single* value as the first argument (in this case, the value of `client`).**

   Note that `switch()` isn't vectorized, so it can't deal with vectors as a first argument.

2. **After the first argument, you give a list of choices with the respected values.**

   Note that you don't have to put quotation marks around the choices.

Remember that `switch()` doesn't work in a vectorized way. You can distinguish the choices more easily, however, so the code becomes more readable.

In fact, the first argument doesn't have to be a value; it can be some expression that evaluates to either a character vector or a number. In case you work with numbers, you don't even have to use `choice=value` in the function call. If you have integers, `switch()` will return the option in that position. In the statement `switch(2, "some value", "something else", "some more")`, the result is `"something else"`. You can find more information and examples on the Help page `?switch`.

Using default values in switch

You don't have to specify all options in a `switch()` call. If you want to have a certain result in case the matched value is not among the specified options, put that result as the last option, without any choice before it. So, the following line of code does exactly the same thing as the nested `ifelse` call from the "Chaining if...else statements" section, earlier in this chapter:

```
VAT <- switch(client, private = 1.12, public = 1.06, 1)
```

You can easily test this out in the console by creating an object called `client` with a certain value and then running the `switch()` call, as in the following example:

```
> client <- "other"
> switch(client, private = 1.12, public = 1.06, 1)
[1] 1
```

To try it yourself, give `client` different values to see how `switch()` works.

Looping Through Values

In the previous section, you used different methods to make choices. Some of these methods aren't vectorized, so you can use only a single value to base your choice on. You could, of course, apply that code on each value you have by hand, but it makes far more sense to automate this task.

Constructing a for loop

As in many other programming languages, you repeat an action for every value in a vector by using a `for` loop. You construct a `for` loop in R as follows:

```
for(i in values){
    . . . do something . . .
}
```

This `for` loop consists of the following parts:

- ✔ The keyword `for`, followed by parentheses.
- ✔ An identifier between the parentheses. In this example, we use `i`, but that can be any object name you like.
- ✔ The keyword `in`, which follows the identifier.
- ✔ A vector with values to loop over. In this example code, we use the object `values`, but that again can be any vector you have available.
- ✔ A code block between braces that has to be carried out for every value in the object `values`.

In the code block, you can use the identifier. Each time R loops through the code, R assigns the next value in the vector `values` to the identifier.

Calculating values in a for loop

Take another look at the priceCalculator() function (refer to the "Making Multiple Choices" section, earlier in this chapter). Earlier, you saw a few possibilities to adapt this function so that you can apply a different VAT rate for public, private, and foreign clients. You can't use any of these options in a vectorized way, but you can use a for loop so the function can calculate the price for multiple clients at once.

Using the values of the vector

Adapt the priceCalculator() function as follows:

```
priceCalculator <- function(hours, pph = 40, client){
    net.price <- hours * pph *
                    ifelse(hours > 100, 0.9, 1)
    VAT <- numeric(0)
    for(i in client){
      VAT <- c(VAT, switch(i, private = 1.12, public = 1.06, 1))
    }

    tot.price <- net.price * VAT
    round(tot.price)
}
```

The first and the last part of the function haven't changed, but in the middle section, you do the following:

1. **Create a numeric vector with length 0 and call it VAT.**

2. **For every value in the vector client, apply switch() to select the correct amount of VAT to be paid.**

3. **In each round through the loop, add the outcome of switch() at the end of the vector VAT.**

The result is a vector VAT that contains, for each client, the correct VAT that needs to be applied. You can test this by adding, for example, a variable type to the data frame clients you created in the previous section. Try this:

```
> clients$type <- c("public", "abroad", "private", "abroad")
> priceCalculator(clients$hours, client = clients$type)
[1] 1060 3960 5040 1600
```

Using loops and indices

The function from the previous section works, but you can write more efficient code if you loop not over the values but over the indices. To do so, replace the middle section in the function with the following code:

```
nclient <- length(client)
VAT <- numeric(nclient)
for(i in seq_along(client)){
  VAT[i] <- switch(client[i], private = 1.12, public = 1.06, 1))
}
```

This code acts very similar to the previous one, but there are a few differences:

✔ You assign the length of the vector `client` to the variable `nclient`.

✔ Then you make a numeric vector `VAT` that is exactly as long as the vector `client`. This is called *pre-allocation* of a vector.

✔ Then you loop over indices of client instead of the vector itself by using the function `seq_along()`. In the first pass through the loop, the first value in `VAT` is set to be the result of `switch()` applied to the first value in `client`. In the second pass, the second value of `VAT` is the result of `switch()` applied to the second value in `client` and so on.

Doing more with loops — and when not to do so

R contains some of the mechanisms used in other programming languages to manipulate loops:

✔ The keyword `next`, to skip to the next iteration of a loop without running the remaining code in the code block

✔ The keyword `break`, to break out of a loop at any given point

✔ The keyword `while`, to construct a loop that continues as long as a certain condition is TRUE

Find more information on the use of these keywords on the Help page `?Control`.

Although you can technically use all three options, they're not often used. Many programmers consider the use of `break` and `next` to be bad coding practice in any language.

For `while`, the situation is a bit more complex. A `while` loop is useful only in very specific cases, like when you generate artificial data that has to meet certain conditions or when you write your own optimization algorithms. But in many cases the built-in optimization functions like `optim()`, `optimize()`, and `nlm()` work faster than a `while` loop — and often give more stable results. These functions require a bit of study before you can apply them, but studying the Help pages `?optim`, `?optimize`, and `?nlm`, as well as related pages, can really pay off.

You may be tempted to replace `seq_along(client)` with the vector `1:nclient`, but that would be a bad idea. If the vector `client` has a length of 0, `seq_along(client)` creates an empty vector and the code in the loop never executes. If you use `1:nclient`, R creates a vector `c(1,0)` and loops over those two values, giving a completely wrong result.

Every time you lengthen an object in R, R copies the whole object and moves it to a new place in memory. This has two effects: First, it slows down your code, because all the copying takes time. Second, as R continuously moves things around in memory, this memory gets split up in a lot of small spaces. This is called *fragmentation,* and it makes the communication between R and the memory less smooth. You can avoid this fragmentation by *pre-allocating* memory as in the previous example.

Looping without Loops: Meeting the Apply Family

Using `for` loops in R has side effects that some programmers would call serious drawbacks. For example, any object that you create or change in a `for` loop is created or changed in the global environment. This may be exactly what you're trying to do, but more often than not, this is an unwanted side effect of the way `for` loops are implemented in R.

Take a look at the following trivial example:

```
> songline <- "Get out of my dreams..."
> for(songline in 1:5) print("...Get into my car!")
```

Contrary to what you may expect, after running this code, the value of `songline` is not the string `"Get out of my dreams..."`, but the number 5, as shown in the output below:

```
> songline
[1] 5
```

Although you never explicitly changed the value of `songline` anywhere in the code, R does so implicitly when carrying out the `for` loop. Every iteration, R reassigns the next value from the vector to `songline` ... in the global environment! By choosing the names of the variables and the identifier wisely, you can avoid running into this kind of trouble. But when writing large scripts, you need to do some serious bookkeeping for the names, and making mistakes becomes all too easy.

To be completely correct, using a `for` loop has an effect on the environment you work in at that moment. If you just use the `for` loop in scripts that you run in the console, the effects will take place in the global environment. If you use a `for` loop in the body of the function, the effects will take place within the environment of that function. For more information, see Chapter 8.

Here's the good news: R has another looping system that's very powerful, that's at least as fast as `for` loops (and sometimes faster), and — most important of all — that doesn't have the side effects of a `for` loop. Actually, this system consists of a complete family of related functions, known as the *apply family*. In base R, this family contains eight functions, all ending with `apply`, and many packages provide additional apply-like functions.

Looking at the family features

Before you start using any of the functions in the apply family, here are the most important properties of these functions:

- ✔ Every one of the apply functions takes at least two arguments: an object and another function. You pass the function as an argument (see Chapter 8).

- ✔ None of these apply functions has side effects. This is the main reason to use them, so we can't stress it enough: If you can use any apply function instead of a `for` loop, use the apply solution. Be aware, though, that possible side effects of the *applied* function are not taken care of by the apply family.

- ✔ Every apply function can pass on arguments to the function that is given as an argument. It does that using the *dots* argument (see Chapter 8).

- ✔ Every function of the apply family always returns a result. Using the apply family makes sense only if you need that result. If you want to print messages to the console with `print()` or `cat()`, for example, there's no point in using the apply family for that.

Meeting three of the members

Say hello to `apply()`, `sapply()`, and `lapply()`, the most used members of the apply family. Each of these functions applies another function to all components in an object. What those components are depends on the object and the apply function you use. Table 9-1 provides an overview of the objects that each of these three functions works on, what each function sees as a component, and which objects each function can return. We explain how to use these functions in the remainder of this chapter.

Table 9-1		Using apply, sapply, and lapply	
Function Name	*Objects the Function Works On*	*What the Function Sees as Components*	*Result Type*
apply	Matrix	Rows or columns	Vector, matrix, array, or list
	Array	Rows, columns, or any dimension	Vector, matrix, array, or list
	Data frame	Rows or columns	Vector, matrix, array, or list
sapply	Vector	Elements	Vector, matrix, or list
	Data frame	Variables	Vector, matrix, or list
	List	Components	Vector, matrix, or list
lapply	Vector	Elements	List
	Data frame	Variables	List
	List	Components	List

Applying functions on rows and columns

In Chapter 7, you calculate the sum of a matrix with the rowSums() function. You can do the same for means with the rowMeans() function, and you have the related functions colSums() and colMeans() to calculate the sum and the mean for each column. But R doesn't have similar functions for every operation you want to carry out. Luckily, you can use the apply() function to apply a function over every row or column of a matrix or data frame.

Counting birds

Imagine you counted the birds in your backyard on three different days and stored the counts in a matrix like this:

```
> counts <- matrix(c(3, 2, 4, 6, 5, 1, 8, 6, 1), ncol = 3)
> colnames(counts) <- c("sparrow", "dove", "crow")
> counts
     sparrow dove crow
[1,]       3    6    8
[2,]       2    5    6
[3,]       4    1    1
```

Each column represents a different species, and each row represents a different day. Now you want to know the maximum count per species on any given day. You could construct a `for` loop to do so, but using `apply()`, you do this in only one line of code:

```
> apply(counts, 2, max)
sparrow   dove   crow
      4      6      8
```

The `apply()` function returns a vector with the maximum for each column and conveniently uses the column names as names for this vector as well. If R doesn't find names for the dimension over which `apply()` runs, it returns an unnamed object instead.

Let's take a look at how this `apply()` function works. In the example you used three arguments:

- ✔ **The object on which the function has to be applied:** In this case, it's the matrix `counts`.

- ✔ **The dimension or index over which the function has to be applied:** The number 1 means row-wise, and the number 2 means column-wise. Here, we apply the function over the columns. In the case of more-dimensional arrays, this index can be larger than 2.

- ✔ **The name of the function that has to be applied:** You can use quotation marks around the function name, but you don't have to. Here, we apply the function `max`. Note that there are no parentheses needed after the function name.

The `apply()` function splits up the matrix (or data frame) in rows (or columns). Remember that if you select a single row or column, R, by default, simplifies that to a vector. The `apply()` function then uses these vectors one by one as an argument to the function you specified. So, the applied function needs to be able to deal with vectors.

Adding extra arguments

Let's go back to our example from the preceding section: Imagine you didn't look for doves the second day. This means that, for that day, you don't have any data, so you have to set that value to `NA` like this:

```
> counts[2, 2] <- NA
```

If you apply the `max()` function on the columns of this matrix, you get the following result:

```
> apply(counts, 2, max)
sparrow   dove   crow
      4     NA      8
```

That's not what you want. In order to deal with the missing values, you need to pass the argument na.rm to the max() function in the apply() call (see Chapter 4). Luckily, this is easily done in R. You just have to add all extra arguments to the function as extra arguments of the apply() call, like this:

```
> apply(counts, 2, max, na.rm = TRUE)
 sparrow    dove    crow
       4       6       8
```

You can pass any arguments you want to the function in the apply() call by just adding them between the parentheses after the first three arguments.

Applying functions to listlike objects

The apply() function works on anything that has dimensions, but what if you don't have dimensions (for example, when you have a list or a vector)? For that, you have two related functions from the apply family at your disposal: sapply() and lapply(). The *l* in lapply stands for list, and the *s* in sapply stands for simplify. The two functions work basically the same — the only difference is that lapply() *always* returns a list with the result, whereas sapply() tries to simplify the final object if possible.

Applying a function to a vector

As you can see in Table 9-1, both sapply() and lapply() consider every value in the vector to be an element on which they can apply a function. Many functions in R work in a vectorized way, so there's often no need to use this.

Using switch on vectors

The switch() function, however, doesn't work in a vectorized way. Consider the following basic example:

```
> sapply(c("a", "b"), switch, a = "Hello", b = "Goodbye")
        a          b
  "Hello" "Goodbye"
```

The sapply() call works very similar to the apply() call from the previous section, although you don't have an argument that specifies the index. Here's a recap:

- ✔ The first argument is the vector on which values you want to apply the function — in this case, the vector c("a", "b").

- ✔ The second argument is the name of the function — in this case, switch.

- ✔ All other arguments are simply the arguments you pass to the switch function.

The sapply() function now takes first the value "a" and then the value "b" as the first argument to switch(), using the arguments a="Hello" and b="Goodbye" each time as the other arguments. It combines both results into a vector and uses the values of c("a", "b") as names for the resulting vector.

The sapply() function has an argument USE.NAMES that you can set to FALSE if you don't want sapply() to use character values as names for the result. For details about this argument, see the Help page ?sapply.

Replacing a complete for loop with a single statement

In the "Calculating values in a for loop" section, earlier in this chapter, you use a for loop to apply the switch() function on all values passed through the argument client. Although that trick works nicely, you can replace the pre-allocation and the loop with one simple statement, like this:

```
priceCalculator <- function(hours, pph = 40, client){
   net.price <- hours * pph * ifelse(hours > 100, 0.9, 1)

   VAT <- sapply(client, switch, private = 1.12, public = 1.06, 1)

   tot.price <- net.price * VAT
   round(tot.price)
}
```

Applying a function to a data frame

You also can use sapply() on lists and data frames. In this case, sapply() applies the specified function on every component in that list. Because data frames are lists as well, everything in this section applies to both lists and data frames.

Imagine that you want to know which type of variables you have in your data frame clients. For a vector, you can use the class() function to find out the type. In order to know this for all variables of the data frame at once, you simply apply the class() function to every variable by using sapply() like this:

```
> sapply(clients, class)
      hours      public       type
  "numeric"   "logical" "character"
```

R returns a named vector that gives you the types of every variable, and it uses the names of the variables as names for the vector. In case you use a named list, R uses the names of the list components as names for the vector.

Simplifying results (or not) with sapply

The `sapply()` function doesn't always return a vector. In fact, the standard output of `sapply` is a list, but that list gets simplified to either a matrix or a vector *if possible*. By default, `sapply()` works in the following way:

- If the result of the applied function on every component of the list or vector is a single number, `sapply()` simplifies the result to a vector.

- If the result of the applied function on every component of the list or vector is a vector with exactly the same length, `sapply()` simplifies the result to a matrix.

- In all other cases, `sapply()` returns a (named) list with the results.

Say you want to know the unique values of every variable in the data frame `clients`. To get all unique values in a vector, you use the `unique()` function. You can get the result you want by applying that function to the data frame `clients` like this:

```
> sapply(clients, unique)
$hours
[1]   25 110 125  40

$public
[1]   TRUE FALSE

$type
[1] "public"  "abroad"  "private"
```

In the variable `hours`, you find four unique values; in the variable `public`, only two; and in the variable `type`, three. Because the lengths of the result differ for every variable, `sapply()` can't simplify the result, so it returns a named list.

Getting lists using lapply

The `lapply()` function works exactly the same as the `sapply()` function, with one important difference: It always returns a list. This trait can be beneficial if you're not sure what the outcome of `sapply()` will be.

Say you want to know the unique values of only a subset of the data frame `clients`. You can get the unique values in the first and third rows of the data frame like this:

```
> sapply(clients[c(1, 3), ], unique)
     hours public  type
[1,] "25"  "TRUE"  "public"
[2,] "125" "FALSE" "private"
```

But because every variable now has two unique values, sapply() simplifies the result to a matrix. If you counted on the result to be a list in the following code, you would get errors. If you used lapply(), on the other hand, you would also get a list in this case, as shown in the following output:

```
> lapply(clients[c(1,3), ], unique)
$hours
[1]   25 125
$public
[1]   TRUE FALSE

$type
[1]  "public"  "private"
```

Actually, the sapply() function has an extra argument, simplify, that you can set to FALSE if you don't want a simplified list, or to the character value "array" if you want to allow simplification to arrays with more than two dimensions. If you set both the arguments simplify and USE.NAMES to FALSE, sapply() and lapply() return exactly the same result. For details on the difference between the two functions, look at the Help file ?sapply.

Chapter 10

Debugging Your Code

*T*o err is human, and programmers fall into that "human" category as well (even though we like to believe otherwise!). Nobody manages to write code without errors, so instead of wondering *if* you have errors in your code, you should ask yourself *where* you have errors in your code. In this chapter, you discover some general strategies and specific tools to find that out.

Knowing What to Look For

A *bug* is simply another word for some kind of error in your program. So, *debugging* doesn't involve insecticides — it just means getting rid of all types of semantic and/or logical errors in your functions.

Before you start hunting down bugs, you have to know what you're looking for. In general, you can divide errors in your code into three different categories:

✔ **Syntax errors:** If you write code that R can't understand, you have syntax errors. Syntax errors always result in an error message and often are caused by misspelling a function or forgetting a bracket.

✔ **Semantic errors:** If you write correct code that doesn't do what you think it does, you have a semantic error. The code itself is correct, but the outcome of that line of code is not. It may, for example, return another type of object than you expect. If you use that object further on, it won't be the type you think it is and your code will fail there.

✔ **Logic errors:** Probably the hardest-to-find are errors in the logic of your code. Your code works, it doesn't generate any errors or warning, but it still doesn't return the result you expect. The mistake is not in the code itself, but in the logic it executes.

This may seem like a small detail, but finding different types of bugs requires different strategies. Often, you can easily locate a syntax error by simply reading the error messages, but semantic errors pose a whole different challenge, and logic errors can hide in your code without your being aware they exist.

Reading Errors and Warnings

If something goes wrong with your code, R tells you. This can happen in two ways:

✔ The code keeps on running until the end, and when the code is finished, R prints out a warning message.

✔ The code stops immediately because R can't carry it out, and R prints out an error message.

We have to admit it: These error messages can range from mildly confusing to completely incomprehensible if you're not used to them. But it doesn't have to stay that way. When you get familiar with the errors and warning messages from R, you can quickly tell what's wrong.

Reading error messages

Let's take a look at such an error message. If you try the following code, you get this more or less clear error message:

```
> "a" + 1
Error in "a" + 1 : non-numeric argument to binary operator
```

You get two bits of information in this error message. First, the line `"a" + 1` tells you in which line of code you have an error. Then it tells you what the error is. In this case, you used a non-numeric argument (the character `"a"`): In combination with a binary operator (the + sign).

R always tells you in which code the error occurs, so you know in many cases where you have to start looking.

Error messages aren't always that clear. Take a look at the following example:

```
> data.frame(1:10, 10:1,)
Error in data.frame(1:10, 10:1, ) : argument is missing, with no default
```

To what argument does this error refer? Actually, it refers to an empty argument you provided for the function. After the second vector, there's a comma that shouldn't be there. A small typing error, but R expects another argument after that comma and doesn't find one.

If you don't immediately understand an error message, take a closer look at the things the error message is talking about. It could be that you simply typed something wrong there.

Caring about warnings (or not)

You can't get around errors, because they just stop your code. Warnings on the other hand are a whole different beast. Even if R throws a warning, it continues to execute the code regardless. So, you can ignore warnings, but in general that's a pretty bad idea. Warnings often are the only sign you have that your code has some semantic or logic error.

For example, you could've forgotten about the ifelse() function discussed in Chapter 9 and tried something like the following example:

```
> x <- 1:10
> y <- if (x < 5 ) 0 else 1
Warning message:
In if (x < 5) 0 else 1 :
  the condition has length > 1 and only the first element will be used
```

This warning points at a semantic error: if() expects a single TRUE or FALSE value, but you provided a whole vector. Note that, just like errors, warnings tell you in general which code has generated the warning.

Here is another warning that pops up regularly and may point to a semantic or logic error in your code:

```
> x <- 4
> sqrt(x - 5)
[1] NaN
Warning message:
In sqrt(x - 5) : NaNs produced
```

Because x - 5 is negative when x is 4, R cannot calculate the square root and warns you that the square root of a negative number is not a number (NaN).

If you're a mathematician, you may point out that the square root of –1 is 0 - 1i. R can, in fact, do calculations on complex numbers, but then you have to define your variables as complex numbers. You can check, for example, ?complex for more information.

Although most warnings result from either semantic or logic errors in your code, even a simple syntax error can generate a warning instead of an error. If you want to plot some points in R, you use the plot() function, as shown in Chapter 16. The function plot() takes an argument col to specify the color of the points, but you could mistakenly try to color the points using the following:

```
> plot(1:10, 10:1, color = "green")
```

If you try this, you get six warning messages at once, all telling you that color is probably not the argument name you were looking for:

```
Warning messages:
1: In plot.window(. . .) : "color" is not a graphical parameter
2: In plot.xy(xy, type, . . .) : "color" is not a graphical parameter
. . . .
```

Notice that the warning messages don't point toward the code you typed at the command line; instead, they point to functions you never used before, like plot.window() and plot.xy(). *Remember:* You can pass arguments from one function to another using the dots argument (see Chapter 8). That's exactly what plot() does here. So, plot() itself doesn't generate a warning, but every function that plot() passes the color argument to does.

If you get warning or error messages, a thorough look at the Help pages of the function(s) that generated the error can help in determining what the reason is for the message you got. For example, at the Help page ?plot.xy, you find that the correct name for the argument is col.

To summarize, most warnings point to one of the following problems:

✔ The function gave a result, but for some reason that result may not be correct.

✔ The function generated an atypical outcome, like NA or NaN values.

✔ The function couldn't deal with some of the arguments and ignored them.

Only the last one tells you there's a problem with your syntax. For the other ones, you have to examine your code a bit more.

Going Bug Hunting

Although the error message always tells you which line of code generates the error, it may not be the line of code where things started going wrong. This makes bug hunting a complex business, but some simple strategies can help you track down these pesky creatures.

Calculating the logit

To illustrate some bug-hunting strategies in R, we use a simple example. Say, for example, your colleague wrote two functions to calculate the logit from both proportions and percentages, but he can't get them to work. So, he asks you to help find the bugs. Here's the code he sends you:

```
# checks input and does logit calculation
logit <- function(x){
  x <- ifelse(x < 0 | x > 1, "NA", x)
  log(x / (1 - x) )
}
# transforms percentage to number and calls logit
logitpercent <- function(x){
  x <- gsub("%", "", x)
  logit(as.numeric(x))
}
```

Type this code into the editor, and save the file using, for example, logitfunc.R as its name. After that, source the file in R from the editor using either the source() function or the source button or command from the editor of your choice. Now the function code is loaded in R, and you're ready to start hunting.

The logit is nothing else but the logarithm of the odds, calculated as log(x / (1-x)) where x is the probability of some event taking place. Statisticians use this when modeling binary data using generalized linear models. If you ever need to calculate a logit yourself, you can use the function qlogis() for that. To calculate probabilities from logit values, you use the plogis() function.

Knowing where an error comes from

Your colleague complained that he got an error with this code:

```
> logitpercent("50%")
Error in 1 - x : non-numeric argument to binary operator
```

Sure enough, but you don't find the code 1 - x in the body of logitpercent(). So, the error comes from somewhere else. To know from where, you can use the traceback() function immediately after the error occurred, like this:

```
> traceback()
2: logit(as.numeric(x)) at logitfunc.R#9
1: logitpercent("50%")
```

This traceback() function prints what is called the *call stack* that led to the last error. This call stack represents the sequence of function calls, but in reverse order. The function at the top is the function in which the actual error is generated.

In this example, R called the logitpercent() function, and that function, in turn, called logit(). The traceback tells you that the error occurred inside the logit() function. Even more, the traceback() function tells you that the error occurred in line 9 of the logitfunc.R code file, as indicated by logitfunc.R#9 in the traceback() output.

The call stack gives you a whole lot of information — sometimes too much. It may point to some obscure internal function as the one that threw the error. If that function doesn't ring a bell, check higher in the call stack for a function you recognize and start debugging from there.

Looking inside a function

Now that you know where the error came from, you can try to find out how the error came about. If you look at the code, you expect that the as.numeric() function in logitpercent() sends a numeric value to the logit() function. So, you want to check what's going on in there.

In ancient times, programmers debugged a function simply by letting it print out the value of variables they were interested in. You can do the same by inserting a few print() statements in the logit() function. This way, you can't examine the object, though, and you have to add and delete print statements at every point where you want to peek inside the function. Luckily, we've passed the age of the dinosaurs; R gives you better methods to check what's going on.

Telling R which function to debug

You can step through a function after you tell R you want to debug it using the debug() function, like this:

```
> debug(logit)
```

From now on, R will switch to the browser mode every time that function is called from anywhere in R, until you tell R explicitly to stop debugging or until you overwrite the function by sourcing it again. To stop debugging a function, you simply use `undebug(logit)`.

If you want to step through a function only once, you can use the function `debugonce()` instead of `debug()`. R will go to browser mode the next time the function is called, and only that time — so you don't need to use `undebug()` to stop debugging.

If you try the function `logitpercent()` again after running the code `debug(logit)`, you see the following in the console:

```
> logitpercent("50%")
debugging in: logit(as.numeric(x))
debug at D:/RForDummies/Ch10/logitfunc.R#2: {
    x <- ifelse(x < 0 | x > 1, "NA", x)
    log(x/(1 - x))
}
Browse[2]>
```

You see that the prompt changed. It now says `Browse[2]`. This prompt tells you that you're browsing inside a function.

The number indicates at which level of the call stack you're browsing at that moment. Remember from the output of the `traceback()` function that the `logit()` function occurred as the second function on the call stack. That's the number 2 in the output above.

The additional text above the changed prompt gives you the following information:

- ✔ The line from where you called the function — in this case, the line `logit(as.numeric(x))` from the `logitpercent()` function
- ✔ The file or function that you debug — in this case, the file `logitfunc.R`, starting at the second line
- ✔ Part of the code you're about to browse through

If you're working in RStudio, you also notice a number of other changes, as shown in Figure 10-1. Right under the environment pane you notice an extra pane called 'Traceback'. It shows you where in the call stack you're currently browsing. Remember that the call stack is the sequence of function calls.

You also see some extra buttons appearing at the top of the console. These buttons are easy shortcuts for the browser commands. Lastly, you see a green arrow in the left margin of your script, indicating at which point in the code you're currently browsing.

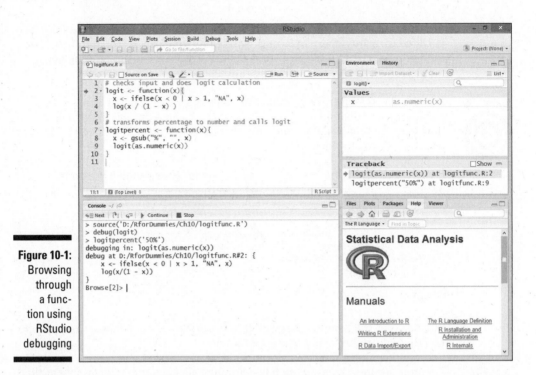

Figure 10-1:
Browsing
through
a func-
tion using
RStudio
debugging

If you didn't save the script file before sourcing, RStudio behaves a bit differently when debugging a function in that script file. RStudio opens a special Source Viewer pane; it shows the code you're debugging, although you can't edit it. RStudio places the green arrow at the place in the code you're currently browsing, but warns you that its guess is approximate because it doesn't have a saved script file or "source" available.

Stepping through the function

When you're in browser mode, you can use any R code you want in order to check the state of different objects. You can browse through the function now with the following commands:

✔ To run the next line of code, type **n** and press Enter. R enters the step-through mode. To run the subsequent lines of code line by line, you don't have to type **n** anymore (although you still can). Just pressing Enter suffices.

✔ To run the remaining part of the code block you're debugging, type **c** and press Enter. Note that when you're in a loop, typing **c** will run the remaining iterations of that loop and drops you off at the first line of code following that loop.

✔ To exit browser mode, type **Q** and press Enter.

You find more browse commands on the help page ?browser. If you want to look at an object that's named like any of the special browse commands, you have to specifically print it out, using either print(n) or str(n).

Try it yourself, by typing **n** in the console, then pressing Enter:

```
Browse[2]> n
debug at D:/RForDummies/Ch10 /logitfunc.R#3: x <- ifelse(x < 0 | x > 1, "NA", x)
```

R now tells you what line it will run next. Because this is the first line in your code, x still has the value that was passed by the logitpercent() function.

It's always smart to check whether that value is what you expect it to be. The logitpercent() function should pass the value 0.50 to logit(), because this is the translation of 50 percent into a proportion. However, if you look at the value of x, you see the following:

```
Browse[2]> str(x)
 num 50
```

Okay, it is a number, but it's 100 times larger than it should be. So, in the logitpercent() function, your colleague made a logical error and forgot to divide by 100. If you correct that in the editor window and then save and source the file again, the test command gives the correct answer:

```
> logitpercent("50%")
[1] 0
```

Start browsing from within the function

This still doesn't explain the error. Your colleague intended to return NA if the number wasn't between 0 and 1, but the function doesn't do that. The number is checked in the ifelse() line in the code, so that is a good place to start looking for an error.

You can easily browse through the logit() function until you reach that point, but when your function is larger, that task can become tedious. R allows you to start the browser at a specific point in your code if you insert a browser() statement at that point. For example, to start the browser mode right after the ifelse() line, you change the body of the logit() function, as in the following code, and source it again:

```
logit <- function(x){
  x <- ifelse(x < 0 | x > 1, "NA", x)
  browser()
  log(x / (1 - x) )
}
```

By sourcing the same function again, you implicitly stop debugging the function. That's why you don't have to un-debug the function explicitly using `undebug(logit)`.

If you now try to run this function again, you see the following:

```
> logit(50)
Called from: logit(50)
Browse[1]>
```

You get less information than you do when you use `debug()`, but you can use the browser mode in exactly the same way as with `debug()`. If you use RStudio, you also notice that RStudio opens up all browser tools, exactly like before.

You can put a `browser()` statement inside a loop as well. If you use the command **c** to run the rest of the code, in this case, R will carry out the function only until the next round in the loop. This way, you can step through the loops of a function.

As you entered the function after the `ifelse()` line, R carried out that code already, so the value of x should be changed to NA. But if you check the value of x now, you see this:

```
Browse[2]> str(x)
  chr "NA"
```

Running the next line finally gives the error. Indeed, your colleague made a semantic error here: He wanted to return NA if the value of x wasn't between 0 and 1, but he accidentally quoted the NA and that makes it a character vector. The code doesn't have any syntax error, but it's still not correct.

If you use `browser()` in your code, don't forget to delete it afterward. Otherwise, your function will continue to switch to browse mode every time you use it. And don't think you won't forget — even experienced R programmers forget this all the time!

Generating Your Own Messages

Generating your own messages may sound strange in a chapter about debugging, but you can prevent bugs by actually generating your own errors. Remember the logic error in the `logitpercent()` function? It would've been easier to spot if the `logit()` function returned an error saying that you passed a number greater than 1.

Adding sensible error (or warning) messages to a function can help debugging future functions where you call that specific function again. It especially helps in finding semantic or logic errors that are otherwise hard to find.

Creating errors

You can tell R to throw an error by inserting the `stop()` function anywhere in the body of the function, as in the following example:

```
logit <- function(x){
  if( any(x < 0 | x > 1) ) stop("x not between 0 and 1")
  log(x / (1 - x) )
}
```

With the `if()` statement, you test whether any value in x lies between 0 and 1. Using the `any()` function around the condition allows your code to work with complete vectors at once, instead of with single values. Because the `log()` function works vectorized as well, the whole function is now vectorized (see Chapter 4).

If you change the body of the `logit()` function this way and try to calculate the logit of 50% and 150% (or 0.5 and 1.5), R throws an error like the following:

```
> logitpercent(c("50%", "150%"))
Error in logit(as.numeric(x)/100) : x not between 0 and 1
```

As the name implies, the execution of the code stops anytime the `stop()` function is actually carried out; hence, it doesn't return a result.

Creating warnings

Your colleague didn't intend for the `logit()` function to stop, though, when some input values were wrong — he just wanted the function to return NA for those values. So, you also could make the function generate a warning instead of an error. That way you still get the same information, but the complete function is carried out so you get a result as well.

To generate a warning, use the `warning()` function instead of the `stop()` function. So, to get the result your colleague wants, you simply change the body of the function to the following code:

```
x <- ifelse(x < 0 | x > 1, NA, x )
if( any(is.na(x)) ) warning("x not between 0 and 1")
log(x / (1 - x) )
```

If you try the function now, you get the desired result:

```
> logitpercent(c("50%", "150%"))
[1]   0 NA
Warning message:
In logit(as.numeric(x)/100) : x not between 0 and 1
```

Not only does the function return NA when it should, but it also gives you a warning that can help with debugging other functions that use the logit() function somewhere in the body.

Recognizing the Mistakes You're Sure to Make

Despite all the debugging tools you have at your disposal, you need some experience to quickly find pesky bugs. But some mistakes are fairly common, and checking whether you made any of these gives you a big chance of pin-pointing the error easily. Some of these mistakes come from default behavior of R you didn't take into account; others are just the result of woolgathering. But every R programmer has made these mistakes at one point, and so will you.

Starting with the wrong data

Probably the most common mistakes in R are made while reading in data from text files using read.table() or read.csv(), as you do in Chapter 12. Many mistakes result in R throwing errors, but sometimes you only notice something went wrong when you look at the structure of your data. In the latter case, you often find that some or all variables are converted to factors when they really shouldn't be (for example, because they should contain only numerical data).

When R gives errors or the structure of your data isn't what you think it should be, check the following:

- **Did you forget to specify the argument header=TRUE?** If so, R will see the column names as values and, as a result, convert every variable to a factor as it always does with character data in a text file.

- **Did you have spaces in your column names or data?** The read.table() function can interpret spaces in, for example, column names or in string data as a separator. You then get errors telling you 'line x did not have y elements'.

✔ **Did you have a different decimal separator?** In some countries, decimals are separated by a comma. You have to specifically tell R that's the case by using the argument `dec=","` in the `read.table()` function.

✔ **Did you forget to specify `stringsAsFactors = FALSE`?** By default, R changes character data to factors, so you always have to add this argument if you want your data to remain character variables.

✔ **Did you have another way of specifying missing values?** R reads `'NA'` in a text file as a missing value, but the file may use a different code (for example, `'missing'`). R will see that as text and again convert that variable to a factor. You solve this by specifying the argument `na.strings` in the `read.table()` function.

If you always check the structure of your data immediately after you read it in, you can catch errors much earlier and avoid hours of frustration. Good practice is to use `str()` for information on the types, and `head()` to see if the values are what you expected.

Having your data in the wrong format

As we've stressed multiple times, every function in R expects your data to be in a specific format. That doesn't mean simply whether it's an integer, character, or factor, but also whether you supply a vector, a matrix, a data frame, or a list. Many functions can deal with multiple formats, but sometimes the result isn't what you expect at all.

In fact, some functions are generic functions that dispatch to a method for the object you supplied as an argument. (See Chapter 8 for more information on dispatching.)

Dropping dimensions when you don't expect it

This mistake is definitely another classic. R automatically tries to reduce the number of dimensions when subsetting a matrix, array, or data frame (see Chapter 7). If you want to calculate the row sums of the numeric variables in a data frame — for example, the built-in data frame `sleep` — you can write a little function like this:

```
rowsum.df <- function(x){
  id <- sapply(x, is.numeric)
  rowSums(x[, id])
}
```

If you try that out on two built-in data frames, `pressure` and `sleep`, you get a result for the first one but the following error message for the second:

```
> rowsum.df(sleep)
Error in rowSums(x[, id]) :
  'x' must be an array of at least two dimensions
```

Because `sleep` contains only a single numeric variable, `x[, id]` returns a vector instead of a data frame, and that causes the error in `rowSums()`.

You can solve this problem either by adding `drop=FALSE` (as shown in Chapter 7) or by using the list subsetting method `x[id]` instead.

Messing up with lists

Although lists help with keeping data together and come in very handy when you're processing multiple datasets, they also can cause some trouble.

First, you can easily forget that some function returns a list instead of a vector. For example, many programmers forget that `strsplit()` returns a list instead of a vector. So, if you want the second word from a sentence, the following code doesn't return an error, but it doesn't give you the right answer either:

```
> strsplit("this is a sentence", " ")[2]
[[1]]
NULL
```

In this example, `strsplit()` returns a list with one component, the vector with the words from the sentence:

```
> strsplit("this is a sentence", " ")
[[1]]
[1] "this"     "is"      "a"        "sentence"
```

To access this vector, you first have to select the wanted component from the list. Only then can you look for the second value using the vector indices:

```
> strsplit("this is a sentence", " ")[[1]][2]
[1] "is"
```

Even the indexing mechanism itself can cause errors of this kind. For example, you have some names of customers and you want to add a dot between their first and last names. So, first, you split them like this:

```
> customer <- c("Johan Delong", "Marie Petit")
> namesplit <- strsplit(customer, " ")
```

You want to paste the second name together with a dot in between, so you need to select the second component from the list. If you use single brackets, you get the following:

```
> paste(namesplit[2], collapse = ".")
[1] "c(\"Marie\", \"Petit\")"
```

That isn't what you want at all. Remember from Chapter 7 that you can use both single brackets and double brackets to select components from a list, but when you use single brackets, you always get a list returned. So, to get the correct result, you need double brackets, like this:

```
> paste(namesplit[[2]], collapse = ".")
[1] "Marie.Petit"
```

Notice that R never gave a sign — not even a warning — that something was wrong. So, if you notice lists where you wouldn't expect them (or don't notice them where you do expect them), check your brackets.

Mixing up factors and numeric vectors

If you work with factors that have numeric values as levels, you have to be extra careful when using these factors in models and other calculations. For example, you convert the number of cylinders in the dataset mtcars to a factor like this:

```
> cyl.factor <- as.factor(mtcars$cyl)
```

If you want to know the median number of cylinders, you may be tempted to do the following:

```
> median(as.numeric(cyl.factor))
[1] 2
```

This result is bogus, because the minimum number of cylinders is four. R converts the internal representation of the factor to numbers, not the labels. So, you get numbers starting from one to the number of levels instead of the original values.

To correctly transform a factor into its original numeric values, you can first transform the factor to character and then to numeric, as shown in Chapter 5. But on very big data, this is done faster with the following construct:

```
> as.numeric(levels(cyl.factor))[cyl.factor]
```

With this code, you create a short vector with the levels as numeric values, and then use the internal integer representation of the factor to select the correct value.

Although R often converts a numeric vector to a factor automatically when necessary, it doesn't do so if both numeric vectors and factors can be used. If you want to model, for example, the mileage of a car to the number of cylinders, you get a different model when you use the number of cylinders as a numeric vector or as a factor. The interpretation of both models is completely different, and a lot depends on what exactly you want to do. But you have to be aware of that, or you may be interpreting the wrong model.

Chapter 11

Getting Help

In This Chapter

▶ Using the built-in R help

▶ Finding information online

▶ Joining the R community

▶ Making a reproducible example to get help

*E*ven the best R programmers occasionally get stuck on a problem. In these situations, you need to know how to find help. Fortunately, R code is generally very well documented and has excellent help available. You just need to know how to access it. In this chapter, we show you how.

If the built-in help doesn't solve your problem, you can search for information on the Internet and turn to the online R community. We end this chapter by walking you through how to create a minimal reproducible example, which you'll find helpful in *getting* help.

Finding Information in the R Help Files

The R documentation (in the form of R Help files) is a rich resource that also can seem cryptic at times. Most of the time, if you read the Help files carefully, you'll get a better understanding of why a function isn't doing what you think it should or why you're getting an error. Some of the R help can look incomprehensible at first glance, but if you persevere — and know what to look for — your investment will pay off.

When you know exactly what you're looking for

If you know the name of the function you need help with, you can access the R Help files in two ways:

- ✔ By typing **help(. . .)** with the function name inside the brackets. For example, typing **help(paste)** returns help about the `paste()` function.

- ✔ By typing **?** followed by the name of the function. For example, typing **?paste** returns help about the `paste()` function.

Typically, the R Help files follow a fairly standard outline. You find most of these sections in every R Help file:

- ✔ **Title:** A one-sentence overview of the function.

- ✔ **Description:** An introduction to the high-level objectives of the function, typically about one paragraph long.

- ✔ **Usage:** A description of the syntax of the function (in other words, how the function is called). This is where you find all the arguments that you can supply to the function, as well as any default values of these arguments.

- ✔ **Arguments:** A description of each argument. Usually this includes a specification of the class (for example, `character`, `numeric`, `list`, and so on). This section is an important one to understand, because supplying an argument of the wrong class is quite frequently a cause of errors in R.

- ✔ **Details:** Extended details about how the function works, provides longer descriptions of the various ways to call the function (if applicable), and a longer discussion of the arguments.

- ✔ **Value:** A description of the class of the value returned by the function.

- ✔ **See also:** Links to other relevant functions. In most of the R editors, you can click these links to read the Help files for these functions.

- ✔ **Examples:** Worked examples of real R code that you can paste into your console and run.

One of the most powerful ways of using R Help is to carefully study the examples in the Examples section. The documentation authors designed these examples to be reproducible, which means that you can copy the whole example to your R console and run it directly. Often, this can help you really understand the nature of the input that each function needs and the output the function gives.

When you don't know exactly what you're looking for

Sometimes you don't know the exact function to use. Or maybe you know the name of the function but you can't remember whether it's spelled in all lowercase letters or with some uppercase letters. In these situations, you have to search the R Help files to find what you need.

You can search the R Help files by typing **help.search(. . .)** with a quoted search term inside the brackets. This gives a list of functions that are similar to the search term; it's useful if you can't remember the exact name of a function, because `help.search()` makes use of fuzzy matching to return a list of matching Help topics. For example, typing **help.search("date")** in the console returns a long list of possible matches, including `format.Date`, `as.POSIXlt`, and `DateTimeClasses`, among others.

Typing two question marks followed by the search term is a shortcut for `help.search()`. For example, typing **??date** returns the same list of functions as typing **help.search("date")** does.

When you search for R help, you get a list of topics that match the search term. For example, you may get this result when typing **??date**:

```
> ??date
ada::update.ada          Add more trees to an ada object
chron::chron             Create a Chronological Object
chron::cut.dates         Create a Factor from a Chron or Dates Object
chron::dates             Generate Dates and Times Components from Input
....
base::Date               Date Class
base::DateTimeClasses    Date-Time Classes
base::diff               Lagged Differences
. . .
```

The left-hand column contains the functions that match your search term, and the right-hand column contains the R Help file title for this function. Notice that each function consists of two elements separated by two colons (for example, `ada::update.ada`). This means, for example, that the package `ada` contains a function called `update.ada()`.

From the description of `update.ada`, it's immediately apparent that this function has nothing to do with dates or times. Nonetheless, it was included in the search results because the function name contained the substring `date`. In this case, if you scroll down the list, you'll also find references to several date functions in the `base` package, including `Date()`, `DateTimeClasses()`, and `diff()`.

After you've identified a function that looks helpful, type **?functionName** to open the relevant Help page. For example, typing **?Date** opens the Help page for Date.

When you use help() or ?, you also can specify a topic name, not just a function name. Some really useful Help pages describe the effect of many functions at once, and because they have a unique topic name, you can access these directly. For example, try reading the Help for ?Syntax, ?Quotes, or ?DateTimeClasses.

Searching the Web for Help with R

Sometimes the built-in R Help simply doesn't give you that moment of inspiration to solve your problem. When this happens, it's time to tap into the information available on the web.

You can search the Internet directly from your R console, by using the RSiteSearch() function. This function enables you to search for keywords in the R documentation, including the help files, vignettes, and task views. RSiteSearch() takes your search term and passes it to the search engine at http://search.r-project.org. Then you can view the search results in your web browser.

For example, to use RSiteSearch() to search for the term *cluster analysis,* use:

```
> RSiteSearch("cluster analysis")
```

Another way of searching the web directly from your console is to use the add-on package called sos and the search function findFn(). This function is a wrapper around RSiteSearch() that combines all the RSiteSearch() results into tabular form on a single page, which may make the results easier to digest.

To use findFn(), you first have to install the sos package:

```
> install.packages("sos")
```

Then you load the package using library("sos"). Finally, you use findFn("cluster"):

```
> library("sos")
> findFn("cluster")
found 2311 matches;  retrieving 20 pages, 400 matches.
2 3 4 5 6 7 8 9 10 11 12 13 14 15 16 17 18 19 20
```

This opens a new tab in your web browser with results in an easy-to-read table. Each row of the table contains a function, the name of the package, and a helpful description and link to the Help page for that function.

If you're trying to search for R topics in your favorite search engine, you may find that the results tend to be unrelated to the programming language. One way of improving the accuracy of your search results is to enclose the *R* in square brackets. For example, to search for the topic of regression in R, use *[R] regression* as your search term. This technique seems to work because the R mailing lists tend to have *[R]* in the topic for each message. In addition, on the Stack Overflow website (`www.stackoverflow.com`), questions that are related to R are tagged with *[r]*.

In addition to your favorite search engine, you also can use the dedicated R search site at `http://search.r-project.org` to search through R functions, vignettes, and the R Help mailing lists. Or you can use the search engine `www.rseek.org`, which is dedicated to R and will search first through all R-related websites for an answer.

Getting Involved in the R Community

Sometimes, no matter how hard you search for help in the mailing list archives, blogs, or other relevant material, you're still stuck. If this ever happens to you, you may want to tap into the R community. R has a very active community made up of people who not only write and share code, but also are very willing to help other R users with their problems.

Discussing R on Stack Overflow and Stack Exchange

Stack Exchange (`www.stackexchange.com`) is a popular website where people ask and answer questions on a variety of topics. It's really a network of sites. Two of the Stack Exchange sites have substantial communities of people asking and answering questions about R:

- **Stack Overflow (`www.stackoverflow.com`):** Here, people discuss programming questions in a variety of programming languages, such as C++, Java, and R.

- **CrossValidated (`http://stats.stackexchange.com`):** Here, people discuss topics related to statistics and data visualization. Because R really excels at these tasks, there is a growing community of R users on CrossValidated.

Both of these sites use tags to identify topics that are discussed. Both sites use the [r] tag to identify questions about R. To find these questions, navigate to the Tags section on the page and type **r** in the search box.

Using the R mailing lists

The R Core Team actively supports four different mailing lists. At www.r-project.org/mail.html, you can find up-to-date information about these lists, as well as find links to subscribe or unsubscribe from the lists. When you subscribe to a mailing list, you can choose to receive either individual email messages or a daily digest.

The four important mailing lists are

- ✔ **R-help:** This is the main R Help mailing list. Anyone can register and post messages on this list, and people discuss a wide variety of topics (for example, how to install packages, how to interpret R's output of statistical results, or what to do in response to warnings and error messages).

- ✔ **R-announce:** This list is for announcements about significant developments in the R code base.

- ✔ **R-packages:** This list is where package authors can announce news about their packages.

- ✔ **R-devel:** This is a specialist mailing list aimed at developers of functions or new R packages — in other words, serious R developers! It's more about programming than about general topics.

Before posting a message to any of the R mailing lists, make sure that you read the posting guidelines, available at www.r-project.org/posting-guide.html. In particular, make sure you include a good, small, reproducible example (see "Making a Minimal Reproducible Example," later in this chapter).

Special interest group mailing lists

In addition to the general mailing lists, you also can participate in about 20 special interest group mailing lists. The R mailing list website www.r-project.org/mail.html also contains links to more than 20 mailing lists for special interest groups.

These special interest groups (SIG) include mailing lists for operating systems, advanced modeling, using R in specific fields, and other specialist development topics. You can also find SIG mailing lists for R-related jobs and teaching.

Tweeting about R

If you want to join the discussion about R on Twitter (www.twitter.com), follow and use the hashtag #rstats. This hashtag attracts discussion from a wide variety of people, including bloggers, package authors, professional R developers, and other interested parties.

Making a Minimal Reproducible Example

When you ask the R community for help, you'll get the most useful advice if you know how to make a minimal reproducible example. A *reproducible example* is a sample of code and data that any other user can run and get the same results as you do. A *minimal* reproducible example is the smallest possible example that illustrates the problem; it consists of

- ✔ A small set of sample data
- ✔ A short snippet of code that reproduces the error
- ✔ The necessary information on your R version, the system it's being run on, and the packages you're using

If you want to know what a minimal reproducible example looks like, take a look at the examples in the R Help files. In general, all the code given in the R Help files fulfills the requirements of a minimal reproducible example.

Creating sample data with random values

In most cases, you can use random data to illustrate a problem. R has some useful built-in functions to generate random numbers and other random data. For example, to make a vector of random numbers, use `rnorm()` for the normal distribution or `runif()` for a uniform distribution. To make a random vector with five elements, try:

```
> set.seed(1)
> x <- rnorm(5)
> x
[1] -0.6264538  0.1836433 -0.8356286  1.5952808  0.3295078
```

You can use the `set.seed()` function to specify a starting seed value for generating random numbers. By setting a seed value, you guarantee that the random numbers are the same each time you run the code. This sounds a bit pointless, doesn't it? It may be pointless in production code, but it's essential for a reproducible example. By setting a seed, you guarantee that your code will produce the same results as another person running your code.

If you want to generate random values of a predetermined set, use the `sample()` function. This function is a bit like dealing from a deck of playing cards. In a card game, you have 52 cards and you know exactly which cards are in the deck. But each deal will be different. You can simulate dealing a hand of seven cards using:

```
> cards <- c(1:9, "J", "Q", "K", "A")
> suits <- c("Spades", "Diamonds", "Hearts", "Clubs")
> deck <- paste(rep(suits, each = 13), cards)
> set.seed(123)
> sample(deck, 7)
[1] "Diamonds 2" "Clubs 2"    "Diamonds 8" "Clubs 5"
[5] "Clubs 7"    "Spades 3"   "Diamonds K"
```

By default, `sample()` uses each value only once. But sometimes you want values to appear multiple times. In this case, you can use the argument `replace=TRUE`. For example, if you want to create a sample of size 12 consisting of the first three letters of the alphabet, use:

```
> set.seed(5)
> sample(LETTERS[1:3], 12, replace = TRUE)
[1] "A" "C" "C" "A" "A" "C" "B" "C" "C" "A" "A" "B"
```

Creating a `data.frame` with sample data is straightforward:

```
> set.seed(42)
> dat <- data.frame(
+     x = sample(1:5),
+     y = sample(c("yes", "no"), 5, replace = TRUE)
+ )
> dat
  x   y
1 5  no
2 4  no
3 1 yes
4 2  no
5 3  no
```

How to use a copy of your own data

Sometimes you have to use a small set of your real-world data in an example. On these occasions, you can first create a subset of your data and then use the function dput() to get an ASCII representation of your data. Then you can paste this representation in your question to the community. As an example, take the built in dataset cars and create an ASCII representation of the first four rows:

```
> dput(cars[1:4, ])
structure(list(speed = c(4, 4, 7, 7),
        dist = c(2, 10, 4, 22)), .Names
        = c("speed",
"dist"), row.names = c(NA, 4L), class =
        "data.frame")
```

Producing minimal code

The hardest part of producing a minimal reproducible example is to keep it minimal. The challenge is to identify the smallest example (the fewest lines of code) that reproduces the problem or error.

Before you submit your code, make sure to describe clearly which packages you use. In other words, remember to include the library() statements. Also, test your code in a new, empty R session to make sure it runs without error. People should be able to just copy and paste your data and your code in the console and get exactly the same results as you get.

Providing the necessary information

Including a little bit of information about your R environment helps people answer your questions. You should consider supplying:

✔ Your R version (for example, R 3.2.0)

✔ Your operating system (for example, Windows 64-bit)

The function sessionInfo() prints information about your version of R and some locale information, as well as attached or loaded packages. Sometimes the output of this function can help you determine whether there are conflicts between your loaded packages. Here's an example of the results of sessionInfo():

```
> sessionInfo()
R version 3.2.0 (2015-04-20)
Platform: x86_64-w64-mingw32/x64 (64-bit)
```

```
locale:
[1] LC_COLLATE=English_United Kingdom.1252
[2] LC_CTYPE=English_United Kingdom.1252
[3] LC_MONETARY=English_United Kingdom.1252
[4] LC_NUMERIC=C
[5] LC_TIME=English_United Kingdom.1252

attached base packages:
[1] stats     graphics  grDevices utils     datasets  methods
[7] base

other attached packages:
[1] ggplot2_1.0.0

loaded via a namespace (and not attached):
 [1] colorspace_1.2-4 digest_0.6.4     grid_3.1.2
 [4] gtable_0.1.2     MASS_7.3-34      munsell_0.4.2
 [7] plyr_1.8.1       proto_0.3-10     Rcpp_0.11.3
[10] reshape2_1.4     scales_0.2.4     stringr_0.6.2
[13] tools_3.1.2
```

The results tell you that this session is running R version 3.2.0 on 64-bit Windows, with a United Kingdom locale. You also can see that R has attached (loaded) the base packages and the add-on package ggplot2 version 1.0.0.

Sometimes it's helpful to include the results of sessionInfo() in your question, because other R users can then tell whether there can be an issue with your R installation.

Part IV
Making the Data Talk

In this part . . .

- ✔ Extract meaningful relationships from data.
- ✔ Subsetting, combining, and restructuring your data.
- ✔ Summarizing data in meaningful ways.
- ✔ Testing your hypotheses.
- ✔ Visit www.dummies.com/extras/r for great Dummies content online.

Chapter 12

Getting Data into and out of R

. . . .

. . . .

*E*very data-processing or analysis problem involves at least three broad steps: input, process, and output. In this chapter, we cover the input and output steps.

Specifically, we look at some of the options you have for importing your data into R, including using the Clipboard, reading data from comma-separated value (CSV) files, and interfacing with spreadsheets like Excel. We also give you some pointers on importing data from commercial statistical software such as SPSS. Next, we give you some options for exporting your data from R. Finally, you manipulate files and folders on your computer.

Getting Data into R

You have several options for importing your data into R, and we cover those options in this section.

Because spreadsheets are so widely used, the bulk of this chapter looks at the different options for importing data originating in spreadsheets. To illustrate the techniques in this chapter, we use a small spreadsheet table with information about the first ten elements of the periodic table, as shown in Figure 12-1.

Figure 12-1:
A spread-sheet with elements of the peri-odic table serves as our example throughout this chapter.

	Atomic number	Name	Symbol	Group	Period	Block	State at STP	Occurrence	Description
1									
2	1	Hydrogen	H	1	1	s	Gas	Primordial	Non-metal
3	2	Helium	He	18	1	s	Gas	Primordial	Noble gas
4	3	Lithium	Li	1	2	s	Solid	Primordial	Alkali metal
5	4	Beryllium	Be	2	2	s	Solid	Primordial	Alkaline earth metal
6	5	Boron	B	13	2	p	Solid	Primordial	Metalloid
7	6	Carbon	C	14	2	p	Solid	Primordial	Non-metal
8	7	Nitrogen	N	15	2	p	Gas	Primordial	Non-metal
9	8	Oxygen	O	16	2	p	Gas	Primordial	Non-metal
10	9	Fluorine	F	17	2	p	Gas	Primordial	Halogen
11	10	Neon	Ne	18	2	p	Gas	Primordial	Noble gas

Entering data in the R text editor

Although R is primarily a programming language, R has a very basic data editor that allows you to enter data directly using the edit() function.

The edit() function is only available in some R code editors, so depending on which software you're using to edit your R code, this approach may not work. The good news is that this option is supported in recent versions of RStudio.

To use the R text editor, first you need to initiate a variable. For example, to create a data frame and manually enter some of the periodic table data, enter the following:

```
> elements <- data.frame()
> elements <- edit(elements)
```

In RStudio, this creates a pop-up window with an interactive editor where you can enter data, as shown in Figure 12-2. Notice that because the data frame is empty, you can scroll left and right, or up and down, to extend the editing range. Notice also that the editor doesn't allow you to modify column or row names.

Enter some data. Then to save your work, click the X in the top-right corner.

Figure 12-2:
Editing data in the R interactive text editor.

Data Editor

File   Edit   Help

	var1	var2	var3	var4
1	1	Hydrogen	H	
2	2	Helium		
3	3	Lithium		
4				

To view the details that you've just entered, use the `print()` function:

```
> print(elements)
  var1     var2 var3
1    1 Hydrogen    H
2    2   Helium   He
3    3  Lithium   Li
```

Using the Clipboard to copy and paste

Another way of importing data interactively into R is to use the Clipboard to copy and paste data.

If you're used to working in spreadsheets and other interactive applications, copying and pasting probably feels natural. If you're a programmer or data analyst, it's much less intuitive. Why? Because data analysts and programmers strive to make their results reproducible. A copy-and-paste action can't be reproduced easily unless you manually repeat the same action. Still, sometimes copying and pasting is useful, so we cover it in this section.

To import data from the Clipboard, use the `readClipboard()` function. For example, select cells B2:B4 in the periodic table spreadsheet, press Ctrl+C to copy those cells to the Clipboard, and then use the following R code:

```
> x <- readClipboard()
> x
[1] "Hydrogen" "Helium"    "Lithium"
```

As you can see, this approach works very well for vector data (in other words, a single column or row of data). But things get just a little bit more complicated when you want to import tabular data to R.

To copy and paste tabular data from a spreadsheet, first select a range in your sheets (for example, cells B1:D5). Then use the `readClipboard()` function and see what happens:

```
> x <- readClipboard()
> x
[1] "Name\tSymbol\tGroup" "Hydrogen\tH\t1"        "Helium\tHe\t1"
[4] "Lithium\tLi\t1"       "Beryllium\tBe\t2"
```

This rather unintelligible result looks like complete gibberish. If you look a little bit closer, though, you'll notice that R has inserted lots instances of `"\t"` into the results. The `"\t"` is the R way of indicating a tab character — in other words, a tab separator between elements of data. Clearly, you need to do a bit more processing on this to get it to work.

The backslash in "\t" is called an *escape sequence*. See the sidebar "Using special characters in escape sequences," later in this chapter, for other examples of frequently used escape sequences in R.

The very powerful read.table() function (which you get to explore in more detail later in this chapter) imports tabular data into R. You can customize the behavior of read.table() by changing its many arguments. Pay special attention to the following arguments:

- ✔ **file:** The name of the file to import. To use the Clipboard, specify file = "clipboard".

- ✔ **sep:** The separator between data elements. In the case of Microsoft Excel spreadsheet data copied from the Clipboard, the separator is a tab, indicated by "\t".

- ✔ **header:** This argument indicates whether the Clipboard data includes a header in the first row (that is, column names). Whether you specify TRUE or FALSE depends on the range of data that you copied.

- ✔ **stringsAsFactors:** If TRUE, this argument converts strings to factors. It's TRUE by default.

Using special characters in escape sequences

Certain keys on your keyboard, such as the Enter and Tab keys, produce behavior without leaving a mark in the document. To use these keys in R, you need to use an *escape sequence* (a special character preceded by a backslash).

Here are some escape sequences you may encounter or want to use:

- ✔ **New line:** \n
- ✔ **Tab stop:** \t
- ✔ **Backslash:** \\
- ✔ **Double quote:** \" . Use this when you need a quote inside a string.
- ✔ **Hexadecimal code:** \xnn

The new line (\n) character comes in handy when you create reports or print messages from your code, while the tab stop (\t) is important when you import some types of delimited text file.

For more information, refer to the section of the R online manual that describes literal constants: http://cran.r-project.org/doc/manuals/R-lang.html#Literal-constants.

```
> x <- read.table(file = "clipboard", sep = "\t", header = TRUE)
> x
      Name Symbol Group
1  Hydrogen      H     1
2    Helium     He     1
3   Lithium     Li     1
4 Beryllium     Be     2
```

Although R offers some interactive facilities to work with data and the Clipboard, it's almost certainly less than ideal for large amounts of data. If you want to import large data files from spreadsheets, you'll be better off using CSV files (described later in this chapter).

Reading data in CSV files

One of the easiest and most reliable ways of getting data into R is to use text files, in particular CSV files. The CSV file format uses commas to separate the different elements in a line, and each line in the textfile represents a single line of data. This makes CSV files ideal for representing tabular data. The additional benefit of CSV files is that almost any data application supports export of data to the CSV format. This is certainly the case for most spreadsheet applications, including Microsoft Excel and LibreOffice Calc.

Some EU countries use an alternative standard where a comma is the decimal separator and a semicolon is the field separator. These settings are stored in a specific set of parameters called the *locale*. When reading or writing CSV files, R ignores those settings; the functions described in this chapter will work the same way on every computer. The same cannot be said about most spreadsheet applications. If you want to open a CSV file generated by R in a spreadsheet, keep in mind that the default decimal and field separators might differ between R and the spreadsheet program you use.

In the following examples, we assume that you have a CSV file stored in a convenient folder in your file system. If you want to reproduce the exact examples, create a small spreadsheet that looks like the example sheet in Figure 12-1. To convert an Excel spreadsheet to CSV format, you need to choose File⇨Save As, which gives you the option to save your file in a variety of formats. Keep in mind that a CSV file can represent only a single worksheet of a spreadsheet. Finally, be sure to use the topmost row of your worksheet (row 1) for the column headings.

An easy way to create this file is to read the `elements` data from the `rfordummies` package (see the sidebar "Getting the `elements` data from the `rfordummies` package").

Using read.csv ( ) to import data

In R, you use the `read.csv()` function to import data in CSV format. This function has a number of arguments, but the only essential argument is `file`, which specifies the location and filename. To read a file called `elements.csv` located in your working directory, use `read.csv()`:

```
> elements <- read.csv("elements.csv")
> str(elements, vec.len = 2)
'data.frame': 118 obs. of  9 variables:
 $ Atomic.no   : int  1 2 3 4 5 . . .
 $ Name        : Factor w/ 118 levels "(Ununhexium) ",..: 50 48 59 16 19 . . .
 $ Symbol      : Factor w/ 118 levels "Ac ","Ag ","Al ",..: 41 42 53 11 9 . . .
 $ Group       : int  1 18 1 2 13 . . .
 $ Period      : int  1 1 2 2 2 . . .
 $ Block       : Factor w/ 4 levels "d ","f ","p ",..: 4 4 4 4 3 . . .
 $ State.at.STP: Factor w/ 4 levels "","Gas ","Liquid ",..: 2 2 4 4 4 . . .
 $ Occurrence  : Factor w/ 3 levels "Primordial ",..: 1 1 1 1 1 . . .
 $ Description : Factor w/ 11 levels "","Actinide",..: 10 9 3 4 8 . . .
```

R imports the data into a data frame. As you can see, this example has 118 observations of 9 variables. Setting the argument vec.len just keeps the output concise for printing in this book. You can safely ignore it.

Notice that the default option is to convert character strings into factors. Thus, the columns Name, Block, State.At.STP, Occurrence, and Description all have been converted to factors. Also, notice that R converts spaces in the column names to periods (for example, in the column State.at.STP).

This default option of converting strings to factors when you use read.table() can be a source of great confusion. Sometimes it can be easier to import data containing strings in such a way that the strings aren't converted factors, but remain character vectors. To import data, keeping the strings as strings, pass the argument stringsAsFactors = FALSE to read.csv() or read.table():

```
> elements <- read.csv("elements.csv", stringsAsFactors = FALSE)
> str(elements, vec.len = 2)
'data.frame': 118 obs. of  9 variables:
 $ Atomic.no   : int  1 2 3 4 5 . . .
 $ Name        : chr  "Hydrogen " "Helium " . . .
 $ Symbol      : chr  "H " "He " . . .
 $ Group       : int  1 18 1 2 13 . . .
 $ Period      : int  1 1 2 2 2 . . .
 $ Block       : chr  "s " "s " . . .
 $ State.at.STP: chr  "Gas " "Gas " . . .
 $ Occurrence  : chr  "Primordial " "Primordial " . . .
 $ Description : chr  "Non-metal" "Noble gas" . . .
```

You can also take more control of the import process by using the argument colClasses. You can pass the colClasses argument to read.csv() to explicitly define whether a specific column should be a numeric, character, or factor. Although this requires you to know the format of your data prior to importing, this small amount of extra work can make it easier to perform your downstream analysis!

If you have a file in the EU format mentioned earlier (where commas are used as decimal separators and semicolons are used as field separators), you need to import it to R using read.csv2() instead of the read.csv() function.

Using read.table () to import tabular data in text files

The CSV format, described in the previous section, is a special case of tabular data in text files. In general, text files can use a multitude of options to distinguish between data elements. For example, instead of using commas, another format is to use tab characters as the separator between columns of data. If you have a tab-delimited file, you can use read.delim() to read your data.

The functions `read.csv()`, `read.csv2()`, and `read.delim()` are special cases of the multipurpose `read.table()` function that can deal with a wide variety of data file formats. The `read.table()` function has a number of arguments that give you fine control over the specification of the text file you want to import. Here are some of these arguments:

✔ **header:** If the file contains column names in the first row, specify TRUE.

✔ **sep:** The data separator (for example, `sep = ","` for CSV files or `sep = "\t"` for tab-separated files).

✔ **quote:** By default, R considers anything between single (`'`) or double (`"`) quotation marks in the text file as a character string. If you want to change that (for example, because an apostrophe in a name is wrongfully seen as the start of a character string), you can change the value of this argument to `quote = "\""`. Note that you have to escape the quotation marks.

✔ **nrows:** If you want to read only a certain number of rows of a file, you can specify this by providing an integer number.

✔ **skip:** Allows you to ignore a certain number of lines before starting to read the rest of the file.

✔ **stringsAsFactors:** If TRUE, it converts strings to factors. It's FALSE by default.

You can access the built-in help by typing **?read.table** into your console.

Using the RStudio data import tool

Although R has an extensive range of functions to import many types of data files, sometimes it's just easier to ask a wizard for help. Recent versions of RStudio have such a tool.

To use the RStudio data import tool, select Tools⇨Import Dataset. From here, you have the option to import either "From Text File" or "From web URL". In both cases, you point to the location of a text file (for example, a comma delimited file).

You then get a nice user interface that allows you to both preview your data and specify the headers, separator, decimal, and quote symbols.

This tool is ideal if you plan to import data only once (or a few times). To automate a regular data import job, you may prefer to write simple scripts using the tools in this chapter.

Reading data from Excel

If you ask users of R what the best way is to import data directly from Microsoft Excel, most of them will probably answer that your best option is to first export from Excel to a CSV file and then use `read.csv()` to import your data to R.

In fact, this is still the advice in Chapter 9 of the R import and export manual, which says, "The first piece of advice is to avoid doing so if possible!" See for yourself at `http://cran.r-project.org/doc/manuals/R-data.html#Reading-Excel-spreadsheets`. The reason is that many of the existing methods for importing data from Excel depend on third-party software or libraries that may be difficult to configure, are not available on all operating systems, or perhaps have restrictive licensing terms.

However, since February 2011, there exists a new alternative: using the package XLConnect, available from CRAN at `http://cran.r-project.org/web/packages/XLConnect/index.html`. What makes XLConnect different is that it uses a Java library to read and write Excel files. This has two advantages:

- **It runs on all operating systems that support Java.** XLConnect is written in Java and runs on Windows, Linux, and Mac OS.

- **There's nothing else to load.** XLConnect doesn't require any other libraries or software. If you have Java installed, it should work.

XLConnect also can write Excel files, including changing cell formatting, in both Excel 97–2003 and Excel 2007/10 formats.

To find out more about XLConnect, you can read the excellent package vignette at `http://cran.r-project.org/web/packages/XLConnect/vignettes/XLConnect.pdf`.

By now you're probably itching to get started with an example. Let's assume you want to read an Excel spreadsheet in your user directory called `Elements.xlsx`. First, install and load the package; then create an object with the filename:

```
> install.packages("XLConnect")
> library("XLConnect")
```

Now you're ready to read a sheet of this workbook with the `readWorksheetFromFile()` function. You need to pass it at least two arguments:

- **file:** A character string with a path to a valid `.xls` or `.xlsx` file

- **sheet:** Either an integer indicating the position of the worksheet (for example, `sheet=1`) or the name of the worksheet (for example, `sheet="Elements"`)

The following two lines do exactly the same thing — they both import the data in the first worksheet (called `Elements`):

```
> elements <- readWorksheetFromFile("Elements.xlsx", sheet = 1)
> elements <- readWorksheetFromFile("Elements.xlsx", sheet = "Elements")
```

Later in this chapter, you learn about a set of functions to do some file manipulation. For now, if you want to remove the CSV and Excel files, and leave your folders in a clean condition, use `file.remove()`:

```
> file.remove(c("elements.xlsx", "elements.csv"))
```

Working with other data types

Despite the fact that CSV files are very widely used to import and export data, they aren't always the most appropriate format. Some data formats allow the specification of data that isn't tabular in nature. Other data formats allow the description of the data using *metadata* (data that describes data).

R includes a *recommended* package called `foreign` with functions to import data files from a number of commercial statistical packages, including SPSS, Stata, SAS, Octave, and Minitab. Table 12-1 lists some of the functions in the `foreign` package.

To use these functions, you first have to load the `foreign` package:

```
> library(foreign)
> read.spss(file = "location/of/myfile")
```

Read the Help documentation on these functions carefully. Because data frames in R may have a quite different structure than datasets in the statistical packages, you have to pay special attention to how value and variable labels are treated by the functions mentioned in Table 12-1. Check also the treatment of special missing values.

Table 12-1	Functions to Import from Commercial Statistical Software Available in the `foreign` Package
System	**Function to Import to R**
SPSS	`read.spss`
SAS	`read.xport` or `read.ssd`
Stata	`read.dta`
Minitab	`read.mtp`

These functions need a specific file format. The function `read.xport()` only works with the XPORT format of SAS. For `read.mtp()`, the file must be in the Minitab portable worksheet (`.mtp`) format.

Note that some of these functions are rather old. The newest versions of the statistical packages mentioned here may have different specifications for the format, so the functions aren't always guaranteed to work.

Note that some of these functions require the statistical package itself to be installed on your computer. For example, `read.ssd()` can work only if you have SAS installed.

The bottom line: If you can transfer data using CSV files, you'll save yourself a lot of trouble.

Finally, if you need to connect R to a database, the odds are that a package exists that can connect to your database of choice. See the sidebar, "Working with databases in R," for some pointers.

Working with databases in R

Data analysts increasingly make use of databases to store large quantities of data or to share data with other people. R has good support to work with a variety of databases, but the exact details of how you do that will vary from system to system.

If you need to connect R to your database, a really good place to start looking for information is in Chapter 4 of the R manual "R data import/export." You can read this chapter at `http://cran.r-project.org/doc/manuals/R-data.html#Relational-databases`.

The package `RODBC` allows you to connect to Open Database Connectivity (ODBC) data sources. You can find this package on CRAN at `http://cran.r-project.org/package=RODBC`.

In addition, you can download and install packages to connect R to many database systems, including:

✔ **MySQL:** The `RMySql` package, available at `http://cran.r-project.org/package=RMySQL`

✔ **SQLite:** The `RSQLite` package, available at `http://cran.r-project.org/package=RSQLite`

✔ **PostgreSQL:** The `RPostgreSQL` package, available at `http://cran.r-project.org/package=RPostgreSQL`

Finally, the `DBI` (Database Interface) package aims to provide a consistent interface to a range of databases. This means you only learn one set of functions to connect and send commands to the database. You can learn more at the CRAN package page at `http://cran.r-project.org/web/packages/DBI/index.html`.

Getting Your Data out of R

For the same reason that it's convenient to import data into R using CSV files, it's also convenient to export results from R to other applications in CSV format. To create a CSV file, use the `write.csv()` function. In the same way that `read.csv()` is a special case of `read.table()`, `write.csv()` is a special case of `write.table()`.

To interactively export data from R for pasting into other applications, you can use `writeClipboard()` or `write.table()`. The `writeClipboard()` function is useful for exporting vector data. For example, to export the names of the built-in dataset `iris`, try the following:

```
> writeClipboard(names(iris))
```

This function doesn't produce any output to the R console, but you can now paste the vector into any application. For example, if you paste this into Excel, you'll have a column of five entries that contains the names of the `iris` data, as shown in Figure 12-3.

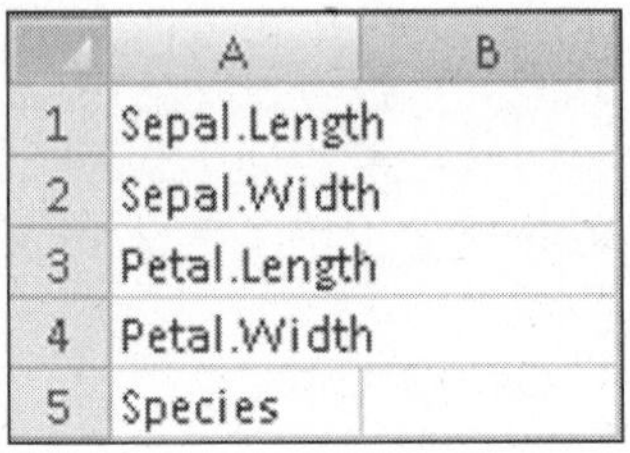

Figure 12-3:
A spreadsheet after first using write-Clipboard() and then pasting.

	A	B
1	Sepal.Length	
2	Sepal.Width	
3	Petal.Length	
4	Petal.Width	
5	Species	

To write tabular data to the Clipboard, you need to use `write.table()` with the arguments `file="clipboard"`, `sep="\t"`, and `row.names=FALSE`:

```
> write.table(head(iris), file = "clipboard", sep = "\t", row.names = FALSE)
```

Again, this doesn't produce output to the R console, but you can paste the data into a spreadsheet. The results look like Figure 12-4.

Figure 12-4:
The first
six lines of
`iris` after
pasting into
a spread-
sheet.

	A	B	C	D	E
1	Sepal.Length	Sepal.Width	Petal.Length	Petal.Width	Species
2	5.1	3.5	1.4	0.2	setosa
3	4.9	3	1.4	0.2	setosa
4	4.7	3.2	1.3	0.2	setosa
5	4.6	3.1	1.5	0.2	setosa
6	5	3.6	1.4	0.2	setosa
7	5.4	3.9	1.7	0.4	setosa

Working with Files and Folders

You know how to import your data into R and export your data from R. Now all you need is an idea of where the files are stored with R and how to manipulate those files.

Understanding the working directory

Every R session has a default location on your operating system's file structure called the *working directory*.

You need to keep track of and deliberately set your working directory in each R session. If you read or write files to disk, this takes place in the working directory. If you don't set the working directory to your desired location, you could easily write files to an undesirable file location.

The `getwd()` function tells you what the current working directory is:

```
> getwd()
[1] "F:/git"
```

To change the working directory, use the `setwd()` function. Be sure to enter the working directory as a character string (enclose it in quotes).

This example shows how to change your working directory to a folder called `F:/git/roxygen2`:

```
> setwd("F:/git/roxygen2")
> getwd()
[1] "F:/git/roxygen2"
```

Notice that the separator between folders is forward slash (/), as it is on Linux and Mac systems. If you use the Windows operating system, the forward slash will look odd, because you're familiar with the backslash (\) of Windows folders. When working in Windows, you need to either use the forward slash or escape your backslashes using a double backslash (\\). Compare the following code:

```
> setwd("F:\\git\\stringr")
> getwd()
[1] "F:/git/stringr"
```

R will always print the results using /, but you're free to use either / or \\ as you please.

To avoid having to deal with escaping backslashes in file paths, you can use the file.path() function to construct file paths that are correct, independent of the operating system you work on. This function is a little bit similar to paste in the sense that it will append character strings, except that the separator is always correct, regardless of the settings in your operating system:

```
> file.path("f:", "git", "surveyor")
[1] "f:/git/surveyor"
```

It's often convenient to use file.path() in setting the working directory. This allows you specify a cascade of drive letters and folder names, and file.path() then assembles these into a proper file path, with the correct separator character:

```
> setwd(file.path("F:", "git", "roxygen2"))
> getwd()
[1] "F:/git/roxygen2"
```

You also can use file.path() to specify file paths that include the filename at the end. Simply add the filename to the path argument. For example, here's the file path to the README.md file in the roxygen2 package installed in a local folder:

```
> file.path("F:", "git", "roxygen2", "roxygen2", "README.md" )
[1] "F:/git/roxygen2/roxygen2/README.md"
```

Manipulating files

Occasionally, you may want to write a script that will traverse a given folder and perform actions on all the files or a subset of files in that folder.

To get a list of files in a specific folder, use list.files() or dir(). These two functions do exactly the same thing, but for backward-compatibility reasons, the same function has two names:

```
> list.files(file.path("F:", "git", "roxygen2"))
[1] "roxygen2"          "roxygen2.Rcheck"
[3] "roxygen2_2.0.tar.gz" "roxygen2_2.1.tar.gz"
```

Table 12-2 lists some other useful functions for working with files.

Next, you get to exercise all your knowledge about working with files. In the next example, you first create a temporary file, then save a copy of the `iris` data frame to this file. To test that the file is on disk, you then read the newly created file to a new variable and inspect this variable. Finally, you delete the temporary file from disk.

Start by using the `tempfile()` function to return a name to a character string with the name of a file in a temporary folder on your system:

```
> my.file <- tempfile()
> my.file
[1] "C:\\Users\\Andrie\\AppData\\Local\\Temp\\tmpGYeLTj\\file14d4366b6095"
```

Notice that the result is purely a character string, not a file. This file doesn't yet exist anywhere. Also note the double \\ as a file separator in Windows, that you encounter in Chapter 5. Remember, the \ is an escape character in R, so to write a backslash, you have to escape it first, hence the \\. If you're on Mac or Linux, R displays the path using the forward slash (/) instead.

Table 12-2	Useful Functions for Manipulating Files
Function	**Description**
list.files	Lists files in a directory.
list.dirs	Lists subdirectories of a directory.
file.exists	Tests whether a specific file exists in a location.
file.create	Creates a file.
file.remove	Deletes files (and directories in Unix operating systems).
tempfile	Returns a name for a temporary file. If you create a file — for example, with `file.create()` or `write.table()` using this returned name — R creates a file in a temporary folder.
tempdir	Returns the file path of a temporary folder on your file system.

You can safely use a forward slash in file paths yourself, even on Windows. R can read file paths using either \\ or /, but it uses the preferred way of your operating system to display a file path.

Next, you save a copy of the data frame iris to my.file using the write.csv() function. Then use list.files() to see if R created the file:

```
> write.csv(iris, file = my.file, , row.names = FALSE)
> list.files(tempdir())
[1] "file14d4366b6095"
```

As you can see, R created the file. Now you can use read.csv() to import the data to a new variable called file.iris:

```
> file.iris <- read.csv(my.file)
```

Use str() to investigate the structure of file.iris. As expected file.iris is a data.frame of 150 observations and six variables. Six variables, you say? Yes, six, although the original iris only has five columns. What happened here was that the default value of the argument row.names of read.csv() is row.names=TRUE. (You can confirm this by taking a close look at the Help for ?read.csv().) So, R saved the original row names of iris to a new column called X:

```
> str(file.iris, vec.len = 2)
'data.frame':  150 obs. of  6 variables:
 $ X           : int  1 2 3 4 5 ...
 $ Sepal.Length: num  5.1 4.9 4.7 4.6 5 ...
 $ Sepal.Width : num  3.5 3 3.2 3.1 3.6 ...
 $ Petal.Length: num  1.4 1.4 1.3 1.5 1.4 ...
 $ Petal.Width : num  0.2 0.2 0.2 0.2 0.2 ...
 $ Species     : Factor w/ 3 levels "setosa","versicolor",..: 1 1 1 1 1 ...
```

To leave your file system in its original order, you can use file.remove() to delete the temporary file:

```
> file.remove(my.file)
> list.files(tempdir())
character(0)
```

As you can see, the result of list.files() is an empty character string, because the file no longer exists in that folder.

Creating reports

Sometimes, the best way of presenting your results is in a nicely formatted report. R has several mechanisms for creating reports, usually by interleaving reporting content and R analysis code in a single document. This style of creating reports is usually referred to as *literate programming*.

For many years, the most widely used scientific literate programming tool that came with R was *Sweave*. Sweave is a format that allows you to embed R into the LaTEX typesetting language, with the resulting output most frequently PDF documents. The name Sweave evokes weaving, where two different types of programming languages (yarn) are interlaced to create a report (cloth or fabric). To learn more, see the Help for ?Sweave.

More recently, rmarkdown, based on knitr has become very popular. The original knitr package allows you to embed R code into a markdown document. This document then is *knitted* to produce HTML output. The word *knit* refers to the process of running the R code, then embedding the results directly inside the text, producing HTML that you can view in a web browser. The rmarkdown package extends knitr to also create PDF or Microsoft Word documents and even HTML presentations. The rmarkdown package is available on CRAN at http://cran.r-project.org/web/packages/rmarkdown/index.html and also is bundled with the RStudio download. A good place to learn more is http://rmarkdown.rstudio.com/.

Chapter 13

Manipulating and Processing Data

• •

In This Chapter

▶ Creating subsets of data

▶ Adding calculated fields

▶ Merging data from different sources

▶ Sorting data

▶ Meeting more members of the apply family

▶ Getting your data into shape

• •

*N*ow it's time to put together all the tools you've seen in earlier chapters. You know how to get data into R, you know how to work with lists and data frames, and you know how to write functions. Combined, these tools form the basic toolkit to be able to do data manipulation and processing in R.

In this chapter, you use some tricks and design idioms for working with data. This includes methods for selecting and ordering data, such as working with lookup tables. You also use some techniques for reshaping data — for example, changing the shape of data from wide format to long format.

Deciding on the Most Appropriate Data Structure

The first decision you have to make before analyzing your data is how to represent that data inside R. In Chapters 4, 5, and 7, you see that the basic data structures in R are vectors, matrices, lists, and data frames.

If your data has only one dimension, then you already know that vectors represent this type of data very well. However, if your data has more than one dimension, you have the choice of using matrices, lists, or data frames. So, the question is: Which do you use when?

Matrices and higher-dimensional arrays are useful when all your data are of a single class — in other words, all your data are numeric or all your data are characters. If you're a mathematician or statistician, you're familiar with matrices and likely use this type of object very frequently.

But in many practical situations, you have data with many different classes — in other words, you have a mixture of numeric and character data. In this case, you need to use either lists or data frames.

If you can imagine your data as a single spreadsheet, a data frame is probably a good choice. Remember that a data frame is simply a list of named vectors of the same length, which is conceptually very similar to a spreadsheet with columns and a column heading for each. If you're familiar with databases, you can think of a data frame as similar to a single table in a database. Data frames are tremendously useful and, in many cases, will be your first choice of object for storing your data.

If your data consists of a collection of objects but you can't represent that as an array or a data frame, then a list is your ideal choice. Because lists can contain all kinds of other objects, including other lists or data frames, they're tremendously flexible. Consequently, R has a wide variety of tools to process lists.

Table 13-1 contains a summary of these choices.

Table 13-1	Useful Objects for Data Analysis	
Object	*Description*	*Comments*
vector	The basic data object in R, consisting of one or more values of a single type (for example, character, number, or integer).	Think of this as a single column or row in a spread-sheet, or a column in a database table.
matrix or array	A multidimensional object of a single type (known as *atomic*). A matrix is an array of two dimensions.	When you have to store numbers in many dimensions, use arrays.
list	Lists can contain objects of any type.	Lists are very useful for storing collections of data that belong together. Lists can contain other lists.
data. frame	Data frames are a special kind of named list where all components have equal length.	Data frames are similar to a single spreadsheet or to a table in a database.

You may find that a data frame is a very suitable choice for most analysis and data-processing tasks. It's a very convenient way of representing your data, and it's similar to working with database tables. When you read data from a comma-separated value (CSV) file with the function `read.csv()` or `read.table()`, R puts the results in a data frame.

Creating Subsets of Your Data

Often the first task in data processing is to create subsets of your data for further analysis. In Chapters 3 and 4, we show you ways of subsetting vectors. In Chapter 7, we outline methods for creating subsets of arrays, data frames, and lists.

Because this is such a fundamental task in data analysis, we review and summarize the different methods for creating subsets of data.

Understanding the three subset operators

You're already familiar with the three subset operators:

- **$:** The dollar-sign operator selects a *single* component of your data (and drops the dimensions of the returned object). When you use this operator with a data frame, the result is always a vector; when you use it with a named list, you get the named component.

- **[[:** The double-square-brackets operator also returns a *single* component, but it offers you the flexibility of referring to the components by position, rather than by name. You use it for data frames and lists.

- **[:** The single-square-brackets operator can return *multiple* components of your data.

Note that `[[` and `[` differ in two important ways: the number of components you can select and the type of object that is returned. Whereas `[[` returns the component itself, `[` returns an object of the same type as the one you subset, dropping of dimensions aside (see Chapter 7 and the section "Subsetting data frames" further in this chapter).

Next, we look at how to use these operators to get exactly the components from your data that you want.

Understanding the five ways of specifying the subset

When you use the single-square-brackets operator, you return multiple components of your data. This means that you need a way of specifying exactly which components you need.

In this paragraph, you subset the built-in dataset `islands`, a named numeric vector with 48 elements:

```
> str(islands)
 Named num [1:48] 11506 5500 16988 2968 16 . . .
 - attr(*, "names")= chr [1:48] "Africa" "Antarctica" "Asia" "Australia" . . .
```

Table 13-2 illustrates the five ways of specifying which components you want to include in or exclude from your data.

Table 13-2	Specifying the Subset Elements	
Subset	*Effect*	*Example*
Blank	Returns all your data	`islands[]`
Positive numerical values	Extracts the elements at these locations	`islands [c(8, 1, 1, 42)]`
Negative numerical values	Extract all but these elements; in other words, excludes these elements	`islands [-(3:46)]`
Logical values	A logical value of TRUE includes element; FALSE excludes element	`islands [islands < 20]`
Character strings	Includes elements where the names match	`islands [c("Madagascar", "Cuba")]`

Subsetting data frames

Having reviewed the rules for creating subsets, you can try it with some data frames. You just have to remember that a data frame is a two-dimensional object and contains rows as well as columns. This means you need to specify the subset for rows and columns independently. To do so, you combine the operators.

To illustrate subsetting of data frames, look at the built-in dataset `iris`, a data frame of 5 columns and 150 rows with data about iris flowers:

```
> str(iris)
'data.frame': 150 obs. of  5 variables:
 $ Sepal.Length: num  5.1 4.9 4.7 4.6 5 5.4 4.6 5 4.4 4.9 . . .
 $ Sepal.Width : num  3.5 3 3.2 3.1 3.6 3.9 3.4 3.4 2.9 3.1 . . .
 $ Petal.Length: num  1.4 1.4 1.3 1.5 1.4 1.7 1.4 1.5 1.4 1.5 . . .
 $ Petal.Width : num  0.2 0.2 0.2 0.2 0.2 0.4 0.3 0.2 0.2 0.1 . . .
 $ Species     : Factor w/ 3 levels "setosa","versicolor",..:
                 1 1 1 1 1 1 1 1 1 1 . . .
```

When you subset objects with more than one dimension, you specify the subset argument for each dimension — you separate the subset arguments with commas.

For example, to get the first five rows of `iris` and all the columns, try:

```
> iris[1:5, ]
```

To get all the rows but only two stated columns, try:

```
> iris[, c("Sepal.Length", "Sepal.Width")]
```

Take special care when subsetting a single column of a data frame, because R may simplify the result. Try:

```
iris[, "Sepal.Length"]
```

Notice that the result is a vector, not a data frame as you would expect.

When your subset operation returns a single column (or vector), the default behavior is to return a simplified version. The way this works is that R inspects the lengths of the returned components. If all these components have the same length, then R simplifies the result to a vector, matrix, or array. In our example, R simplifies the result to a vector. To override this behavior, specify the argument `drop=FALSE` in your subset operation:

```
> iris[, "Sepal.Length", drop = FALSE]
```

Alternatively, you can subset the data frame like a list, as shown in Chapter 7. The following code returns you a data frame with only one column as well:

```
> iris["Sepal.Length"]
```

Finally, to get a subset of only some columns and some rows:

```
> iris[1:5, c("Sepal.Length", "Sepal.Width")]
  Sepal.Length Sepal.Width
1          5.1         3.5
2          4.9         3.0
3          4.7         3.2
4          4.6         3.1
5          5.0         3.6
```

Taking samples from data

Statisticians often have to take samples of data and then calculate statistics. Because a sample is really nothing more than a subset of data, taking a sample is easy with R. To do so, you make use of `sample()`, which takes a vector as input; then you tell it how many samples to draw from that list.

Say you wanted to simulate rolls of a die, and you want to get ten results. Because the outcome of a single roll of a die is a number between one and six, your code looks like this:

```
> sample(1:6, 10, replace = TRUE)
 [1] 2 2 5 3 5 3 5 6 3 5
```

You tell `sample()` to return 10 values, each in the range `1:6`. Because every roll of the die is independent from every other roll of the die, you're sampling with replacement. This means that you take one sample from the list and reset the list to its original state (in other words, you put the element you've just drawn back into the list). To do this, you add the argument `replace=TRUE`, as in the example.

Because the return value of the `sample()` function is a randomly determined number, if you try this function repeatedly, you'll get different results every time. This is the correct behavior in most cases, but sometimes you may want to get repeatable results every time you run the function. Usually, this will occur only when you develop and test your code, or if you want to be certain that someone else can test your code and get the same values you did. In this case, it's customary to specify a so-called *seed value*.

If you provide a seed value, the random-number sequence will be reset to a known state. This is because R doesn't create truly random numbers, but only pseudo-random numbers. A pseudo-random sequence is a set of numbers that, for all practical purposes, seem to be random but were generated by an algorithm. When you set a starting seed for a pseudo-random process, R always returns the same pseudo-random sequence. But if you don't set the seed, R draws from the current state of the random number generator (RNG). On startup, R may set a random seed to initialize the RNG, but each time you call it, R starts from the next value in the RNG stream. You can read the Help for ?RNG to get more detail.

You use the set.seed() function to specify your seed starting value. The argument to set.seed() is any integer value:

```
> set.seed(1)
> sample(1:6, 10, replace = TRUE)
 [1] 2 3 4 6 2 6 6 4 4 1
```

If you draw another sample, without setting a seed, you get a different set of results, as you would expect:

```
> sample(1:6, 10, replace = TRUE)
 [1] 2 2 5 3 5 3 5 6 3 5
```

Now, to demonstrate that set.seed() actually does reset the RNG, try it again. But this time, set the seed once more:

```
> set.seed(1)
> sample(1:6, 10, replace = TRUE)
 [1] 2 3 4 6 2 6 6 4 4 1
```

You get exactly the same results as the first time you used set.seed(1).

You can use sample() to take samples from the data frame iris. In this case, you may want to use the argument replace=FALSE. Because this is the default value of the replace argument, you don't need to write it explicitly:

```
> set.seed(123)
> index <- sample(nrow(iris), 5)
> index
[1]   44 118  61 130 138
> iris[index, ]
    Sepal.Length Sepal.Width Petal.Length Petal.Width    Species
44           5.0         3.5          1.6         0.6     setosa
118          7.7         3.8          6.7         2.2  virginica
61           5.0         2.0          3.5         1.0 versicolor
130          7.2         3.0          5.8         1.6  virginica
138          6.4         3.1          5.5         1.8  virginica
```

Removing duplicate data

A special application of subsetting is finding and removing duplicate values.

The function duplicated() finds duplicate values and returns a logical vector that tells you whether the specific value is a duplicate of a previous value.

This means that for duplicated values, `duplicated()` returns FALSE for the first occurrence and TRUE for every following occurrence of that value, as in the following example:

```
> duplicated(c(1, 2, 1, 3, 1, 4))
[1] FALSE FALSE  TRUE FALSE  TRUE FALSE
```

If you try this on a data frame, R automatically checks the observations (meaning, it treats every row as a value). For example, using `iris`:

```
> duplicated(iris)
  [1] FALSE FALSE FALSE FALSE FALSE FALSE FALSE FALSE FALSE
 [10] FALSE FALSE FALSE FALSE FALSE FALSE FALSE FALSE FALSE
....
[136] FALSE FALSE FALSE FALSE FALSE FALSE FALSE  TRUE FALSE
[145] FALSE FALSE FALSE FALSE FALSE FALSE
```

If you look carefully, you notice that row 143 is a duplicate (because the 143rd value in the resulting vector has the value TRUE). You also can tell this by using the `which()` function:

```
> which(duplicated(iris))
[1] 143
```

To remove the duplicate from `iris`, you need to exclude this row from your data. Remember there are two ways to exclude data using subsetting:

- ✔ **Specify a logical vector, where FALSE means that the respective component of the list will be excluded.** The ! (exclamation point) operator is a logical negation. This means that it converts TRUE into FALSE and vice versa. So, to remove the duplicates from `iris`, you do the following:

  ```
  > iris[!duplicated(iris), ]
  ```

- ✔ **Specify negative values.** In other words:

  ```
  > index <- which(duplicated(iris))
  > iris[-index, ]
  ```

In both cases, you'll notice that your instruction has removed row 143.

Be careful when removing components using negative values. If you create a vector of negative subscripts, you need to make sure it has at least one component. Otherwise you get nothing when you want everything!

Removing rows with missing data

Another application of subsetting data frames is finding and removing rows with missing data. The function to check for this is `complete.cases()`.

Try this on the built-in dataset `airquality`, a data frame with much missing data:

```
> str(airquality)
> complete.cases(airquality)
```

The results of `complete.cases()` is a logical vector with the value `TRUE` for rows that are complete, and `FALSE` for rows that have some `NA` values. To remove the rows with missing data from `airquality`, try the following:

```
> x <- airquality[complete.cases(airquality), ]
> str(x)
```

Your result should be a data frame with 111 rows, rather than the 153 rows of the original `airquality` data frame.

As always with R, there is more than one way of achieving your goal. In this case, you can use `na.omit()` to omit all rows that contain NA values:

```
> x <- na.omit(airquality)
```

When you're certain that your data is clean, you can start to analyze it by adding calculated fields.

If you use any of these methods to subset your data or clean out missing values, remember to store the result in a new object. R doesn't change anything in the original data frame unless you explicitly overwrite it. That's a good thing, because you can't accidently mess up your data.

Adding Calculated Fields to Data

After creating the appropriate subset of your data, the next step in your analysis is to perform some calculations.

Doing arithmetic on columns of a data frame

R makes it very easy to perform calculations on columns of a data frame because each column is itself a vector. This means that you can use all the tools that you encountered in Chapters 4, 5, and 6.

Sticking to the `iris` data frame, try to do a few calculations on the columns. For example, calculate the ratio between the lengths and width of the sepals:

```
> x <- iris$Sepal.Length / iris$Sepal.Width
```

Now you can use all the R tools to examine your result. For example, inspect the first five elements of your results with the `head()` function:

```
> head(x)
[1] 1.457143 1.633333 1.468750 1.483871 1.388889 1.384615
```

As you can see, performing calculations on columns of a data frame is straightforward. Just keep in mind that each column is really a vector, so you simply have to remember how to perform operations on vectors (see Chapter 5).

Using with and transform to improve code readability

After a short while of writing subset statements in R, you'll get tired of typing the dollar sign to extract columns of a data frame. Fortunately, there is a way to reduce the amount of typing and to make your code much more readable at the same time. The trick is to use the `with()` function. Try this:

```
> y <- with(iris, Sepal.Length / Sepal.Width)
```

The `with()` function allows you to refer to columns inside a data frame without explicitly using the dollar sign or even the name of the data frame itself. So, in our example, because you use `with(iris, ...)`, R knows to evaluate both `Sepal.Length` and `Sepal.Width` in the context of `iris`.

Hopefully, you agree that this is much easier to read and understand. With the function `identical()` you can confirm that `y` is in fact identical to `x`:

```
> identical(x, y)
[1] TRUE
```

In addition to `with()`, the helpful `transform()` function allows you to assign values to columns in your data frame very easily. Say you want to add your calculated ratio (sepal length : width) to the original data frame. You're already familiar with writing code like this:

```
> iris$ratio <- iris$Sepal.Length / iris$Sepal.Width
```

Now, using `transform()` it turns into the following:

```
> transform.iris <- transform(iris, ratio = Sepal.Length / Sepal.Width)
```

Now look at the structure of `iris` and notice that `ratio` is a column:

```
> head(transform.iris$ratio)
[1] 1.457143 1.633333 1.468750 1.483871 1.388889 1.384615
```

Note that `with()` and `transform()` are nice helper functions for use when working interactively in the R console. However, these functions make use of some special R magic, called *non-standard evaluation*. Because of the way this evaluation is special, it is generally advised to use `$`, `[` or `[[` in your own scripts, especially inside functions.

Creating subgroups or bins of data

One of the first tasks statisticians use to investigate their data is to draw histograms. (You get to plot histograms in Chapter 15). A *histogram* is a plot of the number of occurrences of data in specific bins or subgroups. Because this type of calculation is fairly common when you do statistics, R has some functions to do exactly that.

The `cut()` function creates bins of equal size (by default) in your data and then classifies each element into its appropriate bin.

If this sounds like a mouthful, don't worry. A few examples should make this come to life.

Using cut to create a fixed number of subgroups

To illustrate the use of `cut()`, have a look at the built-in dataset `state.x77`, an array with several columns and one row for each state in the United States:

```
> head(state.x77)
           Population Income Illiteracy Life Exp Murder HS Grad Frost   Area
Alabama          3615   3624        2.1    69.05   15.1    41.3    20  50708
Alaska            365   6315        1.5    69.31   11.3    66.7   152 566432
Arizona          2212   4530        1.8    70.55    7.8    58.1    15 113417
Arkansas         2110   3378        1.9    70.66   10.1    39.9    65  51945
California      21198   5114        1.1    71.71   10.3    62.6    20 156361
Colorado         2541   4884        0.7    72.06    6.8    63.9   166 103766
```

We want to work with the column called `Frost`. To extract this column, try the following:

```
> frost <- state.x77[, "Frost"]
> head(frost, 5)
   Alabama    Alaska   Arizona  Arkansas California
        20       152        15        65         20
```

You now have a new object, `frost`, a named numeric vector. Now use `cut()` to create three bins in your data:

```
> cut(frost, 3, include.lowest = TRUE)
 [1] [-0.188,62.6] (125,188]     [-0.188,62.6] (62.6,125]
 [5] [-0.188,62.6] (125,188]     (125,188]     (62.6,125]
 ....
[45] (125,188]     (62.6,125]    [-0.188,62.6] (62.6,125]
[49] (125,188]     (125,188]
Levels: [-0.188,62.6] (62.6,125] (125,188]
```

The result is a factor with three levels. The names of the levels seem a bit complicated, but they tell you in mathematical set notation what the boundaries of your bins are. For example, the first bin contains those states that have frost between –0.188 and 62.6 days. In reality, of course, none of the states will have frost on negative days — R is being mathematically conservative and adds a bit of padding.

Note the argument `include.lowest = TRUE` to `cut()`. The default value for this argument is `include.lowest = FALSE`, which can sometimes cause R to ignore the lowest value in your data.

Adding labels to cut

The level names aren't very user-friendly, so specify some better names with the `labels` argument:

```
> cut(frost, 3, include.lowest = TRUE, labels = c("Low", "Med", "High"))
 [1] Low  High Low  Med  Low  High High Med  Low  Low  Low
 ....
[45] High Med  Low  Med  High High
Levels: Low Med High
```

Now you have a factor that classifies states into low, medium, and high, depending on the number of days of frost they get.

Using table to count the number of observations

One interesting piece of analysis is to count how many states are in each bracket. You can do this with the `table()` function, which simply counts the number of observations in each level of your factor:

```
> x <- cut(frost, 3, include.lowest = TRUE, labels = c("Low", "Med", "High"))
> table(x)
x
 Low  Med High
  11   19   20
```

You encounter the `table()` function again in Chapter 15.

Combining and Merging Data Sets

The next thing you may want to do is combine data from different sources. Generally speaking, you can combine different sets of data in three ways:

- ✔ **By adding columns:** If the two sets of data have an equal set of rows, and the order of the rows is identical, then adding columns makes sense. Your options for doing this are `data.frame` or `cbind()` (see Chapter 7).

- ✔ **By adding rows:** If both sets of data have the same columns and you want to add rows to the bottom, use `rbind()` (see Chapter 7).

- ✔ **By combining data with different shapes:** The `merge()` function combines data based on common columns, as well as common rows. In databases language, this is usually called *joining data*.

Figure 13-1 shows these three options schematically.

In this section, we look at some of the possibilities of combining data with `merge()`. More specifically, you use `merge()` to find the intersection, as well as the union, of different data sets. You also look at other ways of working with lookup tables, using the functions `match()` and `%in%`.

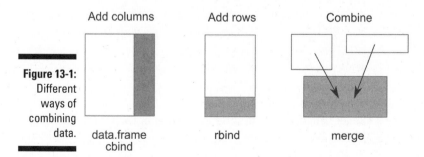

Figure 13-1: Different ways of combining data.

Add columns — data.frame cbind

Add rows — rbind

Combine — merge

Sometimes you want to combine data where it isn't as straightforward to simply add columns or rows. It could be that you want to combine data based on the values of preexisting keys in the data. This is where the merge() function is useful. You can use merge() to combine data only when certain matching conditions are satisfied.

Say, for example, you have information about states in a country. If one dataset contains information about population and another contains information about regions, and both have information about the state name, you can use merge() to combine your results.

Creating sample data to illustrate merging

To illustrate the different ways of using merge, have a look at the built-in dataset state.x77. This is an array, so start by converting it into a data frame. Then add a new column with the names of the states. Finally, remove the old row names. (Because you explicitly add a column with the names of each state, you don't need to have that information duplicated in the row names.)

```
> all.states <- as.data.frame(state.x77)
> all.states$Name <- rownames(state.x77)
> rownames(all.states) <- NULL
```

Now you should have a data frame all.states with 50 observations of nine variables:

```
> str(all.states)
'data.frame':   50 obs. of  9 variables:
 $ Population: num  3615 365 2212 2110 21198 . . .
 $ Income    : num  3624 6315 4530 3378 5114 . . .
 $ Illiteracy: num  2.1 1.5 1.8 1.9 1.1 0.7 1.1 0.9 1.3 2 . . .
 $ Life Exp  : num  69 69.3 70.5 70.7 71.7 . . .
 $ Murder    : num  15.1 11.3 7.8 10.1 10.3 6.8 3.1 6.2 10.7 13.9 . . .
 $ HS Grad   : num  41.3 66.7 58.1 39.9 62.6 63.9 56 54.6 52.6 40.6 . . .
 $ Frost     : num  20 152 15 65 20 166 139 103 11 60 . . .
 $ Area      : num  50708 566432 113417 51945 156361 . . .
 $ Name      : chr  "Alabama" "Alaska" "Arizona" "Arkansas" . . .
```

Creating a subset of cold states

Next, create a subset called cold.states consisting of those states with more than 150 days of frost each year, keeping the columns Name and Frost:

```
> cold.states <- all.states[all.states$Frost>150, c("Name", "Frost")]
> cold.states
```

```
          Name Frost
2        Alaska   152
6      Colorado   166
....
45      Vermont   168
50      Wyoming   173
```

Creating a subset of large states

Finally, create a subset called `large.states` consisting of those states with a land area of more than 100,000 square miles, keeping the columns `Name` and `Area`:

```
> large.states <- all.states[all.states$Area >= 100000, c("Name", "Area")]
> large.states
          Name    Area
2        Alaska 566432
3       Arizona 113417
....
31  New Mexico 121412
43        Texas 262134
```

Now you're ready to explore the different types of merge.

Using the merge() function

In R you use the `merge()` function to combine data frames. This powerful function tries to identify columns or rows that are common between the two different data frames.

Using merge to find the intersection of data

The simplest form of `merge()` finds the intersection between two different sets of data. In other words, to create a data frame that consists of those states that are cold as well as large, use the default version of `merge()`:

```
> merge(cold.states, large.states)
      Name Frost   Area
1   Alaska   152 566432
2 Colorado   166 103766
3  Montana   155 145587
4   Nevada   188 109889
```

If you're familiar with a database language such as SQL, you may have guessed that `merge()` is very similar to a database join. This is, indeed, the case, and the different arguments to `merge()` allow you to perform natural joins, as well as left, right, and full outer joins.

The merge() function takes quite a large number of arguments. These arguments can look quite intimidating until you realize that they form a smaller number of related arguments:

- **x:** A data frame.
- **y:** A data frame.
- **by, by.x, by.y:** The names of the columns that are common to both x and y. The default is to use the columns with common names between the two data frames. The arguments by.x and by.y allow you to state which columns to match even if they don't have the same column name. See the examples in **?merge**
- **all, all.x, all.y:** Logical values that specify the type of merge. The default value is all=FALSE (meaning that only the matching rows are returned).

That last group of arguments — all, all.x and all.y — deserves some explanation. These arguments determine the type of merge that will happen (see the next section).

Understanding the different types of merge

The merge() function allows four ways of combining data:

- **Natural join:** To keep only rows that match from the data frames, specify the argument all=FALSE.
- **Full outer join:** To keep all rows from both data frames, specify all=TRUE.
- **Left outer join:** To include all the rows of your data frame x and only those from y that match, specify all.x=TRUE.
- **Right outer join:** To include all the rows of your data frame y and only those from x that match, specify all.y=TRUE.

You can see a visual depiction of all these different options in Figure 13-2.

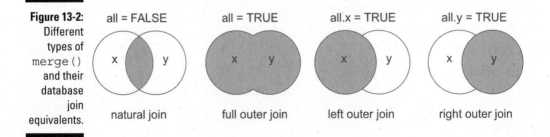

Figure 13-2: Different types of merge() and their database join equivalents.

all = FALSE — natural join
all = TRUE — full outer join
all.x = TRUE — left outer join
all.y = TRUE — right outer join

Finding the union (full outer join)

Returning to the examples of U.S. states, to perform a complete merge of cold and large states, use merge() and specify all=TRUE:

```
> merge(cold.states, large.states, all = TRUE)
        Name Frost   Area
1      Alaska   152 566432
2     Arizona    NA 113417
3  California    NA 156361
....
13      Texas    NA 262134
14    Vermont   168     NA
15    Wyoming   173     NA
```

Both data frames have a variable Name, so R matches the cases based on the names of the states. The variable Frost comes from the data frame cold.states, and the variable Area comes from the data frame large.states.

Note that this performs the complete merge and fills the columns with NA values where there is no matching data.

Working with lookup tables

Sometimes doing a full merge of the data isn't exactly what you want. In these cases, it may be more appropriate to match values in a lookup table. To do this, you can use the match() or %in% functions.

Finding a match

The match() function returns the matching positions of two vectors or, more specifically, the positions of first matches of one vector in the second vector. For example, to find which large states also occur in the data frame cold.states, you can do the following:

```
> index <- match(cold.states$Name, large.states$Name)
> index
 [1]  1  4 NA NA  5  6 NA NA NA NA NA
```

As you see, the result is a vector that indicates matches were found at positions one, four, five, and six. You can use this result as an index to find all the large states that are also cold states.

Keep in mind that you need to remove the NA values first, using na.omit():

```
> large.states[na.omit(index), ]
        Name   Area
2      Alaska 566432
6    Colorado 103766
26    Montana 145587
28     Nevada 109889
```

Making sense of %in%

A very convenient alternative to match() is the function %in%, which returns a logical vector indicating whether there is a match.

The %in% function is a special type of function called a *binary operator*. This means you use it by placing it between two vectors, unlike most other functions where the arguments are in parentheses:

```
> index <- cold.states$Name %in% large.states$Name
> index
 [1]  TRUE  TRUE FALSE FALSE  TRUE  TRUE FALSE FALSE FALSE FALSE FALSE
```

If you compare this to the result of match(), you see that you have a TRUE value for every non-missing value in the result of match(). Or, to put it in R code, the operator %in% does the same as the following code:

```
> !is.na(match(cold.states$Name, large.states$Name))
 [1]  TRUE  TRUE FALSE FALSE  TRUE  TRUE FALSE FALSE FALSE FALSE
```

The match() function returns the indices of the matches in the *second* argument for the values in the *first* argument. On the other hand, %in% returns TRUE for every value in the *first* argument that matches a value in the *second* argument. The order of the arguments is important here.

Because %in% returns a logical vector, you can use it directly to subset values in a vector.

```
> cold.states[index, ]
        Name Frost
2      Alaska   152
6    Colorado   166
26    Montana   155
28     Nevada   188
```

The %in% function is an example of a binary operator in R. This means that the function is used by putting it between two values, as you would for other operators, such as + (plus) and - (minus). At the same time, %in% is in *infix operator*. An infix operator in R is identifiable by the percent signs around the function name. If you want to know how %in% is defined, look at the details section of its Help page. But note that you have to place quotation marks around the function name to get the Help page, like this: ?"%in%".

Sorting and Ordering Data

Another common task in data analysis and reporting is to sort information. You can answer many everyday questions with *league tables* — sorted tables that tell you the best or worst of specific things. For example, parents want to know which school in their area is the best, and businesses need to know the most productive factories or the most lucrative sales areas. When you have the data, you can answer all these questions simply by sorting it.

As an example, look again at the built-in data about the states in the U.S. First, create a data frame called some.states that contains information contained in the built-in variables state.region and state.x77:

```
> some.states <- data.frame(
+     Region = state.region,
+     state.x77)
```

To keep the example manageable, create a subset of only the first ten rows and the first three columns:

```
> some.states <- some.states[1:10, 1:3]
> some.states
              Region Population Income
Alabama        South       3615   3624
Alaska          West        365   6315
Arizona         West       2212   4530
....
Delaware       South        579   4809
Florida        South       8277   4815
Georgia        South       4931   4091
```

You now have a variable called some.states that is a data frame consisting of ten rows and three columns (Region, Population, and Income).

Sorting vectors

R makes it easy to sort vectors in either ascending or descending order. Because each column of a data frame is a vector, you may find that you perform this operation quite frequently.

Sorting a vector in ascending order

To sort a vector, you use the sort() function. For example, to sort Population in ascending order, try this:

```
> sort(some.states$Population)
 [1]   365   579  2110  2212  2541  3100  3615  4931  8277
[10] 21198
```

Sorting a vector in decreasing order

You also can tell sort() to go about its business in decreasing order. To do this, specify the argument decreasing=TRUE:

```
> sort(some.states$Population, decreasing = TRUE)
 [1] 21198  8277  4931  3615  3100  2541  2212  2110   579
[10]   365
```

You can access the Help documentation for the sort() function by typing ?sort into the R console.

Sorting data frames

Another way of sorting data is to determine the order that elements should be in, if you were to sort. This sounds long-winded, but as you'll see, having this flexibility means you can write statements that are very natural.

Getting the order

First, determine the element order to sort state.info$Population in ascending order. Do this using the order() function:

```
> order.pop <- order(some.states$Population)
> order.pop
 [1]  2  8  4  3  6  7  1 10  9  5
```

This means to sort the elements in ascending order, you first take the second element, then the eighth element, then the fourth element, and so on. Try it:

```
> some.states$Population[order.pop]
 [1]   365   579  2110  2212  2541  3100  3615  4931  8277
[10] 21198
```

Yes, this is rather long-winded. But next we look at how you can use order() in a very powerful way to sort a data frame.

Sorting a data frame in ascending order

In the preceding section, you calculated the order in which the elements of `Population` should be in order for it to be sorted in ascending order, and you stored that result in `order.pop`. Now, use `order.pop` to sort the data frame `some.states` in ascending order of population:

```
> some.states[order.pop, ]
              Region Population Income
Alaska          West        365   6315
Delaware       South        579   4809
Arkansas       South       2110   3378
....
Georgia        South       4931   4091
Florida        South       8277   4815
California      West      21198   5114
```

Sorting in decreasing order

Just like `sort()`, the `order()` function also takes an argument called `decreasing`. For example, to sort `some.states` in decreasing order of population:

```
> order(some.states$Population)
 [1]  2  8  4  3  6  7  1 10  9  5
> order(some.states$Population, decreasing = TRUE)
 [1]  5  9 10  1  7  6  3  4  8  2
```

Just as before, you can sort the data frame `some.states` in decreasing order of population. Try it, but this time don't assign the order to a temporary variable:

```
> some.states[order(some.states$Population, decreasing = TRUE), ]
              Region Population Income
California      West      21198   5114
Florida        South       8277   4815
Georgia        South       4931   4091
....
Arkansas       South       2110   3378
Delaware       South        579   4809
Alaska          West        365   6315
```

Sorting on more than one column

You probably think that sorting is very straightforward, and you're correct. Sorting on more than one column is almost as easy.

You can pass more than one vector as an argument to the `order()` function. If you do so, the result will be the equivalent of adding a secondary sorting key. In other words, the order will be determined by the first vector and any ties will then sort according to the second vector.

Next, you sort `some.states` on more than one column — in this case, `Region` and `Population`. If this sounds confusing, don't worry — it really isn't. Try it yourself. First, calculate the order to sort `some.states` by region as well as population:

```
> index <- with(some.states, order(Region, Population))
> some.states[index, ]
                    Region Population Income
Connecticut Northeast         3100   5348
Delaware          South        579   4809
Arkansas          South       2110   3378
Alabama           South       3615   3624
Georgia           South       4931   4091
Florida           South       8277   4815
Alaska            West         365   6315
Arizona           West        2212   4530
Colorado          West        2541   4884
California        West       21198   5114
```

Sorting multiple columns in mixed order

You may start to wonder how to calculate the order when some of the columns need to be in increasing order and others need to be in decreasing order.

To do this, you need to make use of a helper function called `xtfrm()`. This function transforms a vector into a numeric vector that sorts in the same order. After you've transformed a vector, you can take the negative to indicate decreasing order.

To sort `some.states` into decreasing order of region and increasing order of population, try the following:

```
> index <- order(-xtfrm(some.
      states$Region),
+     some.states$Population)
> some.states[index, ]
                    Region Population Income
Alaska            West         365   6315
Arizona           West        2212   4530
Colorado          West        2541   4884
California        West       21198   5114
Delaware          South        579   4809
Arkansas          South       2110   3378
Alabama           South       3615   3624
Georgia           South       4931   4091
Florida           South       8277   4815
Connecticut Northeast         3100   5348
```

Traversing Your Data with the Apply Functions

R has a powerful suite of functions that allows you to apply a function repeatedly over the components of a list. The interesting and crucial thing about this is that it happens without an explicit loop. In Chapter 9, you see how to use loops appropriately and get a brief introduction to the apply family.

Because this is such a useful concept, you'll come across quite a few different flavors of functions in the apply family of functions. The specific flavor of `apply()` depends on the structure of data that you want to traverse:

✔ **Array or matrix:** Use the `apply()` function. This traverses either the rows or columns of a matrix, applies a function to each resulting vector, and returns a vector (or array or list) of summarized results.

✔ **List:** Use the `lapply()` function to traverse a list, apply a function to each component, and return a list of the results. Sometimes it's possible to simplify the resulting list into an array, matrix, or vector. This is what the `sapply()` function does.

Figure 13-3 demonstrates the appropriate function, depending on whether your data is in the form of an array or a list.

The ability to apply a function over the components of a list is one of the distinguishing features of the functional programming style as opposed to an imperative programming style. In the imperative style, you use loops, but in the functional programming style you apply functions. R has a variety of apply-type functions, including `apply()`, `lapply()`, and `sapply()`.

Array   List

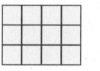

Figure 13-3:
Use `apply` on arrays and matrices; use `lapply` or `sapply` on lists and data frames.

Rows: apply(x, 1, FUN)

Columns: apply(x, 2, FUN)

lapply(x, FUN, ...) produces list

sapply(x, FUN, ...) produces vector or array, if possible

Using the apply() function to summarize arrays

If you have data in the form of an array or matrix and you want to summarize this data, the `apply()` function is really useful. The `apply()` function traverses an array or matrix by column or row and applies a summarizing function.

The `apply()` function takes four arguments:

- ✔ **X:** This is your data — an array (or matrix).

- ✔ **MARGIN:** A numeric vector indicating the dimension over which to traverse; 1 means rows and 2 means columns.

- ✔ **FUN:** The function to apply (for example, `sum` or `mean`).

- ✔ **. . . (dots):** If your `FUN` function requires any additional arguments, you can add them here.

To illustrate this, look at the built-in dataset `Titanic`. This is a four-dimensional table with passenger data of the ship *Titanic,* describing their cabin class, gender, age, and whether they survived.

```
> str(Titanic)
 table [1:4, 1:2, 1:2, 1:2] 0 0 35 0 0 0 17 0 118 154 . . .
 - attr(*, "dimnames")=List of 4
 ..$ Class   : chr [1:4] "1st" "2nd" "3rd" "Crew"
 ..$ Sex     : chr [1:2] "Male" "Female"
 ..$ Age     : chr [1:2] "Child" "Adult"
 ..$ Survived: chr [1:2] "No" "Yes"
```

You can treat a table as an array. Under the hood, a table is an array of integer values. In Chapter 14, you find some more examples of working with tables, including information differences between tables and arrays.

To find out how many passengers were in each of their cabin classes, you need to summarize `Titanic` over its first dimension, `Class`:

```
> apply(Titanic, 1, sum)
 1st  2nd  3rd Crew
 325  285  706  885
```

Similarly, to calculate the number of passengers in the different age groups, you need to apply the `sum()` function over the third dimension:

```
> apply(Titanic, 3, sum)
Child Adult
  109  2092
```

You also can apply a function over two dimensions at the same time. To do this, you need to combine the desired dimensions with the `c()` function. For example, to get a summary of how many people in each age group survived, you do the following:

```
> apply(Titanic, c(3, 4), sum)
        Survived
Age        No Yes
  Child    52  57
  Adult  1438 654
```

Using lapply() and sapply() to traverse a list or data frame

In Chapter 9 we show you how to use the `lapply()` and `sapply()` functions. In this section, we briefly review these functions.

When your data is in a list or data frame, and you want to perform calculations on each component of that list, the appropriate `apply` function is `lapply()`. For example, to get the class of each component of `iris`, try:

```
> lapply(iris, class)
```

As you know, when you use `sapply()`, R attempts to simplify the results to a matrix or vector:

```
> sapply(iris, class)
Sepal.Length  Sepal.Width Petal.Length  Petal.Width      Species
   "numeric"    "numeric"    "numeric"    "numeric"     "factor"
```

Say you want to calculate the mean of each column of `iris`:

```
> sapply(iris, mean)
Sepal.Length  Sepal.Width Petal.Length  Petal.Width      Species
    5.843333     3.057333     3.758000     1.199333           NA
Warning message:
In mean.default(X[[5L]], . . .) :
  argument is not numeric or logical: returning NA
```

There is a problem with this line of code. It throws a warning message because `species` is not a numeric column. So, you may want to write a small function inside `sapply()` that tests whether the argument is numeric. If it is, then calculate the mean score; otherwise, simply return NA.

In Chapter 8, you create your own functions. The `FUN` argument of the `apply()` functions can be any function, including your own custom functions. In fact, you can go one step further. It's actually possible to define a function *inside* the `FUN` argument call to any `apply()` function:

```
> sapply(iris, function(x) if(is.numeric(x)) mean(x) else NA)
Sepal.Length  Sepal.Width Petal.Length  Petal.Width      Species
    5.843333     3.057333     3.758000     1.199333           NA
```

What's happening here? You defined a function that takes a single argument `x`. If `x` is numeric, it returns `mean(x)`; otherwise, it returns `NA`. Because `sapply()` traverses your list, each column, in turn, is passed to your function and evaluated.

When you define a nameless function like this inside another function, it's called an *anonymous function*. Anonymous functions are useful when you want to calculate something fairly simple, but you don't necessarily want to permanently store that function in the global environment.

Using tapply() to create tabular summaries

So far, you've used three members of the apply family of functions: `apply()`, `lapply()`, and `sapply()`. It's time to meet the fourth member of the family. You use `tapply()` to create tabular summaries of data. This function takes three arguments:

- ✔ **X:** A vector
- ✔ **INDEX:** A factor or list of factors
- ✔ **FUN:** A function

With `tapply()`, you can easily create summaries of subgroups in data. For example, calculate the mean sepal length in the dataset `iris`:

```
> tapply(iris$Sepal.Length, iris$Species, mean)
    setosa versicolor  virginica
     5.006      5.936      6.588
```

With this short line of code, you do some powerful stuff. You tell R to take the `Sepal.Length` column, split it according to `Species`, and then calculate the mean for each group.

This is an important idiom for writing code in R, and it usually goes by the name Split, Apply, and Combine (SAC). In this case, you split a vector into groups, apply a function to each group, and then combine the result into a vector.

Of course, using the `with()` function, you can write your line of code in a slightly more readable way:

```
> with(iris, tapply(Sepal.Length, Species, mean))
    setosa versicolor  virginica
     5.006     5.936      6.588
```

Using `tapply()`, you also can create more complex tables to summarize your data. You do this by using a list as your `INDEX` argument.

Using tapply() to create higher-dimensional tables

For example, try to summarize the data frame `mtcars`, a built-in data frame with data about motor-car engines and performance. As with any object, you can use `str()` to inspect its structure:

```
> str(mtcars)
```

The variable `am` is a numeric vector that indicates whether the engine has an automatic (`0`) or manual (`1`) gearbox. Because this isn't very descriptive, start by creating a new object, `cars`, that is a copy of `mtcars`, and change the column `am` to be a factor:

```
> cars <- transform(mtcars,
+     am = factor(am, levels = 0:1, labels = c("Automatic", "Manual"))
+ )
```

Now use `tapply()` to find the mean miles per gallon (`mpg`) for each type of gearbox:

```
> with(cars, tapply(mpg, am, mean))
Automatic    Manual
 17.14737  24.39231
```

Yes, you're correct. This is still only a one-dimensional table. Now, try to make a two-dimensional table with the type of gearbox (`am`) and number of gears (`gear`):

```
> with(cars, tapply(mpg, list(gear, am), mean))
  Automatic Manual
3  16.10667     NA
4  21.05000 26.275
5       NA 21.380
```

You use `tapply()` to create tabular summaries of data. This is a little bit similar to the `table()` function. However, `table()` can create only contingency tables (that is, tables of counts), whereas with `tapply()`, you can specify any function as the aggregation function. In other words, with `tapply()`, you can calculate counts, means, or any other value.

If you want to summarize statistics on a single vector, `tapply()` is very useful and quick to use.

Using aggregate ( )

Another R function that does something very similar is `aggregate()`:

```
> with(cars, aggregate(mpg, list(gear = gear, am = am), mean))
  gear       am        x
1    3 Automatic 16.10667
2    4 Automatic 21.05000
3    4    Manual 26.27500
4    5    Manual 21.38000
```

Next, you take `aggregate()` to new heights using the formula interface.

Getting to Know the Formula Interface

Now it's time to get familiar with another very important idea in R: the formula interface. The formula interface allows you to concisely specify which columns to use when fitting a model, as well as the behavior of the model.

It's important to keep in mind that the formula notation refers to statistical formulae, as opposed to mathematical formulae. So, for example, the formula operator + means to include a column, not to mathematically add two columns together. Table 13-3 contains some formula operators, as well as examples and their meanings. You need these operators when you start building models.

We won't go deeper into this subject in this book, but now you know what to look for in the Help pages of different modeling functions. Be aware of the fact that the interpretation of the signs can differ depending on the modeling function you use.

Many R functions allow you to use the formula interface, often in addition to other ways of working with that function. For example, the `aggregate()` function also allows you to use formulae:

```
> aggregate(mpg ~ gear + am, data = cars, mean)
  gear       am      mpg
1    3 Automatic 16.10667
2    4 Automatic 21.05000
3    4    Manual 26.27500
4    5    Manual 21.38000
```

Table 13-3	Some Formula Operators and Their Meanings	
Operator	*Example*	*Meaning*
~	y ~ x	Model y as a function of x
+	y ~ a + b	Include columns a as well as b
-	y ~ a - b	Include a but exclude b
:	y ~ a : b	Estimate the interaction of a and b
*	y ~ a * b	Include columns as well as their interaction (that is, y ~ a + b + a:b)
|	y ~ a | b	Estimate y as a function of a conditional on b

Notice that the first argument is a formula and the second argument is the source data frame. In this case, you tell aggregate() to model mpg as a function of gear and am, calculating the mean. By using the formula interface your function becomes very easy to read. Notice also the last column is correctly named mpg, not x, as it was with the default method, so your output is nicely formatted and easier to work with.

When you look at the Help file for a function, it'll always be clear whether you can use a formula with that function. For example, take a look at the Help for ?aggregate. In the usage section of this page, you find the following text:

```
## S3 method for class 'data.frame'
aggregate(x, by, FUN, . . ., simplify = TRUE)

## S3 method for class 'formula'
aggregate(formula, data, FUN, . . .,
          subset, na.action = na.omit)
```

This page lists a method for class data.frame, as well as a method for class formula. This indicates that you can use either formulation.

You can find more (technical) information about formula on its own Help page, ?formula.

In the next section, we offer yet another example of using the formula interface for reshaping data.

Some more uses of formula

In Chapter 15, you get to do some statistical modeling using R. In particular, you use `aov()` to do an analysis of variance (ANOVA).

To do an ANOVA using the same data, try the following:

```
> aov(mpg ~ gear + am, data = cars)
```

To find out how to interpret the results of ANOVA, turn to Chapter 15.

You can use formulae to specify your model in just about every R statistical modeling function, such as ANOVA and linear regression.

Another use of the formula interface is in graphics, especially in the package `lattice` that you get to use in Chapter 17. To plot the data in our example, try this:

```
> library("lattice")
> xyplot(mpg ~ gear + am, data = cars)
```

Whipping Your Data into Shape

Often, a data analysis task boils down to creating tables with summary information, such as aggregated totals, counts, or averages. Say, for example, you have information about four games of Granny, Geraldine, and Gertrude:

```
  Game  Venue Granny Geraldine Gertrude
1  1st Bruges     12         5       11
2  2nd  Ghent      4         4        5
3  3rd  Ghent      5         2        6
4  4th Bruges      6         4        7
```

You now want to analyze the data and get a summary of the total scores for each player in each venue:

```
   variable Bruges Ghent
1    Granny     18     9
2 Geraldine      9     6
3  Gertrude     18    11
```

If you use spreadsheets, you may be familiar with the term *pivot table*. The functionality in pivot tables is essentially the ability to group and aggregate data and to perform calculations.

In the world of R, people usually refer to this as the process of reshaping data. In base R, there is a function, `reshape()`, that does this, but we discuss how to use the add-on package `reshape2`, which you can find on CRAN.

Understanding data in long and wide formats

When talking about reshaping data, it's important to recognize data in long and wide formats. These visual metaphors describe two ways of representing the same information.

You can recognize data in wide format by the fact that columns generally represent groups. So, our example of basketball games is in wide format, because there is a column for the baskets made by each of the participants:

```
  Game  Venue Granny Geraldine Gertrude
1  1st  Bruges    12        5       11
2  2nd  Ghent      4        4        5
3  3rd  Ghent      5        2        6
4  4th  Bruges     6        4        7
```

In contrast, have a look at the long format of exactly the same data:

```
    Game   Venue   variable value
1    1st  Bruges     Granny    12
2    2nd   Ghent     Granny     4
3    3rd   Ghent     Granny     5
4    4th  Bruges     Granny     6
5    1st  Bruges  Geraldine     5
6    2nd   Ghent  Geraldine     4
7    3rd   Ghent  Geraldine     2
8    4th  Bruges  Geraldine     4
9    1st  Bruges   Gertrude    11
10   2nd   Ghent   Gertrude     5
11   3rd   Ghent   Gertrude     6
12   4th  Bruges   Gertrude     7
```

Notice how, in the long format, the three columns for Granny, Geraldine, and Gertrude have disappeared. In their place, you now have a column called `value` that contains the actual score, and a column called `variable` that links the score to either of the three ladies.

When converting data between long and wide formats, it's important to be able to distinguish identifier variables from measured variables:

- ✔ **Identifier variables:** Identifier, or ID, variables identify the observations. Think of these as the key that identifies your observations. (In database design, these are called primary or secondary keys.)
- ✔ **Measured variables:** This represents the measurements you observed.

In our example, the identifier variables are `Game` and `Venue`, while the measured variables are the goals (that is, the columns `Granny`, `Geraldine`, and `Gertrude`).

Getting started with the reshape2 package

Base R has a function, `reshape()`, that works fine for data reshaping. However, the original author of this function had in mind a specific use case for reshaping: so-called longitudinal data.

Longitudinal research takes repeated observations of a research subject over a period of time. For this reason, longitudinal data typically has the variables associated with time.

The problem of data reshaping is far more generic than simply dealing with longitudinal data. For this reason, Hadley Wickham wrote and released the package `reshape2` that contains several functions to convert data between long and wide format.

To download and install `reshape2`, use `install.packages()`:

```
> install.packages("reshape2")
```

At the start of each new R session that uses `reshape2`, you need to load the package into memory using `library()`:

```
> library("reshape2")
```

Now you can start. First, create some data:

```
> goals <- data.frame(
+    Game = c("1st", "2nd", "3rd", "4th"),
+    Venue = c("Bruges", "Ghent", "Ghent", "Bruges"),
+    Granny = c(12, 4, 5, 6),
+    Geraldine = c(5, 4, 2, 4),
+    Gertrude = c(11, 5, 6, 7)
+ )
```

This constructs a wide data frame with five columns and four rows with the scores of Granny, Geraldine, and Gertrude.

Melting data to long format

You've already seen the words *wide* and *long* as visual metaphors for the shape of your data. In other words, wide data tends to have more columns and fewer rows compared to long data. The reshape package extends this metaphor by using the terminology of *melt* and *cast:*

- ✔ To convert wide data to long, you melt it with the melt() function.
- ✔ To convert long data to wide, you cast it with the dcast() function for data frames or the acast() function for arrays.

Try converting your wide data frame goals to a long data frame using melt():

```
> mgoals <- melt(goals)
Using Game, Venue as id variables
```

The melt() function tries to guess your identifier variables (id.vars), if you don't provide them explicitly, and tells you which ones it used. By default, melt() considers all categorical variables (factors) as identifier variables. This is often a good guess, and exactly what you want in this example.

Specifying your identifier variables explicitly is a good idea. You do this by adding an argument id.vars, specifying the column names of the identifiers:

```
> mgoals <- melt(goals, id.vars = c("Game", "Venue"))
```

The new object, mgoals, now contains your data in long format:

```
> mgoals
    Game  Venue  variable value
1    1st Bruges    Granny    12
2    2nd  Ghent    Granny     4
3    3rd  Ghent    Granny     5
. . .
10   2nd  Ghent  Gertrude     5
11   3rd  Ghent  Gertrude     6
12   4th Bruges  Gertrude     7
```

Casting data to wide format

Now that you have a *molten dataset* (a dataset in long format), you're ready to reshape it. To illustrate that the process of reshaping keeps all your data intact, try to reconstruct the original:

```
> dcast(mgoals, Venue + Game ~ variable, sum)
    Game  Venue Granny Geraldine Gertrude
1    1st Bruges     12         5       11
2    2nd  Ghent      4         4        5
3    3rd  Ghent      5         2        6
4    4th Bruges      6         4        7
```

Can you see how `dcast()` takes a formula as its second argument? More about that in a minute, but first inspect your results. It should match the original data frame.

Next, you may want to do something more interesting — for example, create a summary by venue and player.

You use the `dcast()` function to cast a molten data frame. To be clear, you use this to convert from a long format to a wide format, but you also can use this to aggregate into intermediate formats, similar to the way a pivot table works.

The `dcast()` function takes three arguments:

- ✔ **data:** A molten data frame.

- ✔ **formula:** A formula that specifies how you want to cast the data. This formula takes the form `x_variable ~ y_variable`. But we simplified it to make a point. You can use multiple *x*-variables, multiple *y*-variables, and even *z*-variables. We say more about that in a few paragraphs.

- ✔ **fun.aggregate:** A function to use if the casting formula results in data aggregation (for example, `length()`, `sum()`, or `mean()`).

So, to get that summary of venue versus player, you need to use `dcast()` with a casting formula `variable ~ Venue`. Note that the casting formula refers to columns in your molten data frame:

```
> dcast(mgoals, variable ~ Venue , sum)
   variable Bruges Ghent
1    Granny     18     9
2 Geraldine      9     6
3  Gertrude     18    11
```

If you want to get a table with the venue running down the rows and the player across the columns, your casting formula should be `Venue ~ variable`:

```
> dcast(mgoals,  Venue ~ variable , sum)
  Venue Granny Geraldine Gertrude
1 Bruges     18         9       18
2  Ghent      9         6       11
```

It's actually possible to have more complicated casting formulae. According to the Help page for `dcast()`, the casting formula takes this format:

```
x_variable + x_2 ~ y_variable + y_2 ~ z_variable ~ . . .
```

Notice that you can combine several variables in each dimension with the plus sign (+), and you separate each dimension with a tilde (~). Also, if you have two or more tildes in the formula (that is, you include a *z*-variable), your result will be a multidimensional array.

So, to get a summary of goals by `Venue`, player (`variable`), and `Game`, you do the following:

```
> dcast(mgoals, Venue + variable ~ Game , sum)
  Venue   variable 1st 2nd 3rd 4th
1 Bruges    Granny  12   0   0   6
2 Bruges Geraldine   5   0   0   4
3 Bruges  Gertrude  11   0   0   7
4  Ghent    Granny   0   4   5   0
5  Ghent Geraldine   0   4   2   0
6  Ghent  Gertrude   0   5   6   0
```

One of the reasons you should understand data in long format is that both the graphics packages `lattice` (Chapter 17) and `ggplot2` (Chapter 18) make extensive use of long format data. The benefit is that you can easily create plots of your data that compare different subgroups. For example, the following code generates Figure 13-4:

```
> library("ggplot2")
> ggplot(mgoals, aes(x = variable, y = value, fill = Game)) +
+   geom_bar(stat = "identity")
```

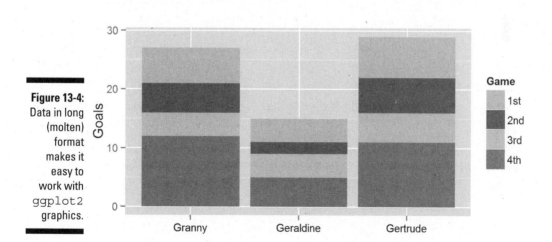

Figure 13-4: Data in long (molten) format makes it easy to work with `ggplot2` graphics.

Chapter 14

Summarizing Data

. . . .

In This Chapter

▶ Using statistical measures to describe your variables

▶ Using convenience functions to summarize variables and data frames

▶ Comparing two groups

. . . .

*I*t's time to get down to the core business of R: statistics! Because R is designed to do just that, you can apply most common statistical techniques with a single command. In general, these commands are very well documented both in the Help files and on the web. If you need more advanced methods or to implement cutting-edge research, very often there's a package for that, and many of these packages come with a book filled with examples.

It may seem rather odd that it takes us 14 chapters to come to the core business of R. Well, it isn't. The difficult part is very often getting the data in the right format. After you've done that, R allows you to carry out the planned analyses rather easily. To see just how easy it is, read on.

R allows you to do just about anything you want, even if the analysis you carry out doesn't make sense at all. R gives you the correct calculation, but that's not necessarily the right answer to your question. In fact, R is like a professional workbench available for everybody. If you don't know what you're doing, chances are, things will get bloody at some point. So, make sure you know the background of the tests you apply on your data, or look for guidance by a professional. All techniques you use in this chapter are explained in the book *Statistics For Dummies,* 2nd Edition, by Deborah J. Rumsey, PhD (Wiley).

Starting with the Right Data

Before you attempt to describe your data, you have to make sure your data is in the right format. This means

✔ Making sure all your data is contained in a data frame (or in a vector if it's a single variable)

✔ Ensuring that all the variables are of the correct type

✔ Checking that the values are all processed correctly

The previous chapters give you a whole set of tools for doing exactly these things. We can't stress enough how important this is. Many of the mistakes in data analysis originate from wrongly formatted data.

Using factors or numeric data

Some data can have only a limited number of different values. For example, people can be either male or female, and you can describe most hair types with only a few colors. Sometimes more values are theoretically possible but not realistic. For example, cars can have more than 16 cylinders in their engines, but you won't find many of them. In one way or another, all this data can be seen as *categorical*. By this definition, categorical data also includes ordinal data (see Chapter 5).

On the other hand, you have data that can have an unlimited amount of possible values. This doesn't necessarily mean that the values can be any value you like. For example, the mileage of a car is expressed in miles per gallon, often rounded to the whole mile. Yet, the real value will be slightly different for every car. The only thing that defines how many possible values you allow is the precision with which you express the data. Data that can be expressed with any chosen level of precision is *continuous*. Both the interval-scaled data and the ratio-scaled data described in Chapter 5 are usually continuous data.

The distinction between categorical and continuous data isn't always clear though. Age is, in essence, a continuous variable, but it's often expressed in the number of years since birth. You still have a lot of possible values if you do that, but what happens if you look at the age of the kids at your local high school? Suddenly you have only five, maybe six, different values in your data. At that point, you may get more out of your analysis if you treat that data as categorical.

When describing your data, you need to make the distinction between data that benefits from being converted to a factor and data that needs to stay numeric. If you can view your data as categorical, converting it to a factor helps with analyzing it.

Counting unique values

Let's take another look at the dataset `mtcars`. This built-in dataset describes fuel consumption and ten different design points from 32 cars from the 1970s. It contains, in total, 11 variables, but all of them are numeric. Although you can work with the data frame as is, some variables could be converted to a factor because they have a limited amount of values.

If you don't know how many different values a variable has, you can get this information in two simple steps:

1. **Get the unique values of the variable using** `unique()`.

2. **Get the length of the resulting vector using** `length()`.

Using the `sapply()` function from Chapter 9, you can do this for the whole data frame at once. You apply an anonymous function combining both mentioned steps on the whole data frame, like this:

```
> sapply(mtcars, function(x) length(unique(x)))
 mpg cyl disp   hp drat   wt qsec   vs   am gear carb
  25   3   27   22   22   29   30    2    2    3    6
```

So, it looks like the variables `cyl`, `vs`, `am`, `gear`, and `carb` can benefit from a conversion to factor. ***Remember:*** You have 32 different observations in that dataset, so all the variables have duplicates for at least some of their values.

When to treat a variable like a factor depends a bit on the situation, but, as a general rule, avoid more than ten different levels in a factor and try to have at least five values per level.

Preparing the data

In many real-life cases, you get heaps of data in a big file, and often in a format you can't use at all. That must be the golden rule of data gathering: Make sure your statistician sweats his pants off just by looking at the data. But no worries! With R at your fingertips, you can quickly shape your data exactly as you want it. Selecting only the variables you need and transforming them to the right format becomes pretty easy with the tricks you see in the previous chapters.

Let's prepare the data frame `mtcars` using some simple tricks. First, create a data frame `cars` using the `transform()` function:

```
> cars <- transform(mtcars[c(1, 2, 9, 10)],
+     gear = ordered(gear),
+     am = factor(am, labels = c("auto", "manual")))
```

With this code, you do the following:

- ✔ **Select four variables from the data frame** `mtcars` **and save them in a data frame called** `cars`. Note that you use the index system for lists to select the variables (see Chapter 7).
- ✔ **Make the variable** `gear` **in this data frame an ordered factor.**
- ✔ **Give the variable** `am` **the value** `"auto"` **if its original value is** `0`, **and** `"manual"` **if its original value is** `1`.
- ✔ **Transform the new variable** `am` **to a factor.**

After running this code, you should have a dataset `cars` in your workspace with the following structure:

```
> str(cars, vec.len = 2)
'data.frame':   32 obs. of  4 variables:
 $ mpg : num  21 21 22.8 21.4 18.7 . . .
 $ cyl : num  6 6 4 6 8 . . .
 $ am  : Factor w/ 2 levels "auto","manual": 1 1 1 2 2 . . .
 $ gear: Ord.factor w/ 3 levels "3"<"4"<"5": 2 2 2 1 1 . . .
```

With this dataset in your workspace, you're ready to tackle the rest of this chapter.

In order to avoid too much clutter on the screen, we set the argument `vec.len=2` in the `str()` function when creating the output. This argument defines the default number of values that are displayed for each variable. If you use `str(cars)`, your output may look a bit different from the one shown here. See the Help page `?str` for more information. Or just forget about it — you'll never use it unless you start writing a book about R.

Describing Continuous Variables

You have the dataset and you've formatted it to fit your needs, so now you're ready for the real work. Analyzing your data always starts with describing it. This way you can detect errors in the data, and you can decide which models are appropriate to get the information you need from the data you have. Which descriptive statistics you use depends on the nature of your data, of course. Let's first take a look at some things you want to do with continuous data.

Talking about the center of your data

Sometimes you're more interested in the general picture of your data than you are in the individual values. You may be interested not in the mileage of every car, but in the average mileage of all cars from that dataset. For this, you calculate the mean using the `mean()` function, like this:

```
> mean(cars$mpg)
[1] 20.09062
```

You also could calculate the average number of cylinders those cars have, but this doesn't really make sense. The average would be 6.1875 cylinders, and we have yet to see a car driving with an incomplete cylinder. In this case, the *median* — the most central value in your data — makes more sense. You get the median from using the function `median()`, like this:

```
> median(cars$cyl)
[1] 6
```

There are numerous other reasons for calculating the median instead of the mean, or even both together. Both statistics describe a different property of your data, and even the combination can tell you something. If you don't know how to interpret these statistics, *Statistics For Dummies,* 2nd Edition, by Deborah J. Rumsey, PhD (Wiley) is a great resource.

Describing the variation

A single number doesn't tell you that much about your data. Often it's at least as important to have an idea about the spread of your data. You can look at this spread using a number of different approaches.

First, you can calculate either the *variance* or the *standard deviation* to summarize the spread in a single number. For that, you have the convenient functions `var()` for the variance and `sd()` for the standard deviation. For example, you calculate the standard deviation of the variable `mpg` in the data frame `cars` like this:

```
> sd(cars$mpg)
[1] 6.026948
```

Checking the quantiles

In addition to the mean and variation, you also can take a look at the quantiles. A *quantile,* or percentile, tells you how much of your data lies below a certain value. The 50 percent quantile, for example, is the same as the median. Again, R has some convenient functions to help you with looking at the quantiles.

Calculating the range

The most-used quantiles are actually the 0 percent and 100 percent quantiles. You could just as easily call them the minimum and maximum, because that's what they are. We introduce the `min()` and `max()` functions in Chapter 4. You can get both together using the `range()` function. This function conveniently gives you the range of the data. So, to know the range of mileages, you simply do:

```
> range(cars$mpg)
[1] 10.4 33.9
```

Calculating the quartiles

The range still gives you only limited information. Often statisticians report the first and the third *quartile* together with the range and the median. These quartiles are, respectively, the 25 percent and 75 percent quantiles, which are the numbers for which one-fourth and three-fourths of the data is smaller. You get these numbers using the `quantile()` function, like this:

```
> quantile(cars$mpg)
    0%    25%    50%    75%   100%
10.400 15.425 19.200 22.800 33.900
```

The quartiles are not the same as the lower and upper hinge calculated in the five-number summary. The latter two are, respectively, the median of the lower and upper half of your data, and they differ slightly from the first and third quartiles. To get the five number statistics, you use the `fivenum()` function.

Getting on speed with the quantile function

The `quantile()` function can give you any quantile you want. For that, you use the `probs` argument. You give the `probs` (or probabilities) as a fractional number. For the 20 percent quantile, for example, you use `0.20` as an argument for the value. This argument also takes a vector as a value, so you can, for example, get the 5 percent and 95 percent quantiles like this:

```
> quantile(cars$mpg, probs = c(0.05, 0.95))
    5%    95%
11.995 31.300
```

The default value for the `probs` argument is a vector representing the minimum (0), the first quartile (0.25), the median (0.5), the third quartile (0.75), and the maximum (1).

All functions from the previous sections have an argument `na.rm` that allows you to remove all `NA` values before calculating the respective statistic. If you don't do this, any vector containing `NA` will have `NA` as a result. This works identically to the `na.rm` argument of the `sum()` function (see Chapter 4).

Describing Categories

A first step in every analysis consists of calculating the descriptive statistics for your dataset. You have to get to know the data you received before you can accurately decide what models you try out on them. You need to know something about the range of the values in your data, how these values are distributed in the range, and how values in different variables relate to each other. Much of what you do and how you do it depends on the type of data.

Counting appearances

Whenever you have a limited number of different values, you can get a quick summary of the data by calculating a *frequency table*. A frequency table is a table that represents the number of occurrences of every unique value in the variable. In R, you use the `table()` function for that.

Creating a table

You can tabulate, for example, the amount of cars with a manual and an automatic gearbox using the following command:

```
> amtable <- table(cars$am)
> amtable

 auto manual
   13     19
```

This outcome tells you that your data contains 13 cars with an automatic gearbox and 19 with a manual gearbox.

Working with tables

As with most functions, you can save the output of `table()` in a new object (in this case, called `amtable`). At first sight, the output of `table()` looks like a named vector, but is it?

```
> class(amtable)
[1] "table"
```

The `table()` function generates an object of the class `table`. These objects have the same structure as an array. Arrays can have an arbitrary number of dimensions and dimension names (see Chapter 7). Tables can be treated as arrays to select values or dimension names.

In the "Describing Multiple Variables" section, later in this chapter, you use multidimensional tables and calculate margins and proportions based on those tables.

Calculating proportions

After you have the table with the counts, you can easily calculate the proportion of each count to the total simply by dividing the table by the total counts. To calculate the proportion of manual and automatic gearboxes in the dataset `cars`, you can use the following code:

```
> amtable / sum(amtable)

   auto   manual
0.40625 0.59375
```

Yet, R also provides the `prop.table()` function to do the same. You can get the exact same result as the previous line of code by doing the following:

```
> prop.table(amtable)
```

You may wonder why you would use an extra function for something that's as easy as dividing by the sum. The `prop.table()` function also can calculate marginal proportions (see the "Describing Multiple Variables" section, later in this chapter).

Finding the center

In statistics, the *mode* of a categorical variable is the value that occurs most frequently. It isn't exactly the center of your data, but if there's no order in your data — if you look at a nominal variable — you can't really talk about a center either.

Although there isn't a specific function to calculate the mode, you can get it by combining a few tricks:

1. **To get the counts for each value, use `table()`.**

2. **To find the location of the maximum number of counts, use `max()`.**

3. **To find the mode of your variable, select the name corresponding with the location in Step 2 from the table in Step 1.**

So, to find the mode for the variable `am` in the dataset `cars`, you can use the following code:

```
> id <- amtable == max(amtable)
> names(amtable)[id]
[1] "manual"
```

The variable `id` contains a logical vector that has the value `TRUE` for every value in the table `amtable` that is equal to the maximum in that table. You select the name from the values in `amtable` using this logical vector as an index.

You also can use the `which.max()` function to find the location of the maximum in a vector. This function has one important disadvantage, though: If you have multiple maxima, `which.max()` returns the position of the first maximum only. If you're interested in all maxima, you should use the construct in the previous example. Also, note that the R function `mode()` does exist, but this mode refers to the storage type of the object, not the measure of centrality.

Describing Distributions

Sometimes the information about the center of the data just isn't enough. You get some information about your data from the variance or the quantiles, but still you may miss important features of your data. Instead of calculating yet more numbers, R offers you some graphical tools to inspect your data further. And in the meantime, you can impress people with some fancy plots.

Plotting histograms

To get a clearer idea about how your data is distributed within the range, you can plot a histogram. In Chapter 16, you fancy up your plots, but for now let's just check the most-used tool for describing your data graphically.

Making the plot

To make a histogram for the mileage data, you simply use the `hist()` function, like this:

```
> hist(cars$mpg, col = "grey")
```

The result of this function is shown on the left of Figure 14-1. There you see that the `hist()` function first cuts the range of the data in a number of even intervals, and then counts the number of observations in each interval. The bar height is proportional to those frequencies. On the *y*-axis, you find the counts.

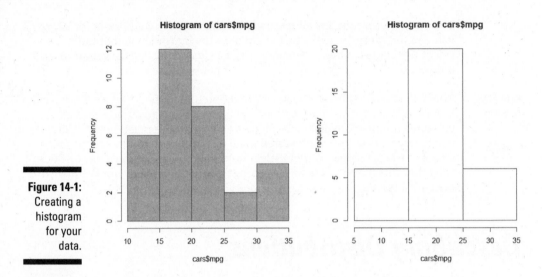

Figure 14-1:
Creating a
histogram
for your
data.

With the argument `col`, you give the bars in the histogram a bit of color.
In Chapter 16, we give you some more tricks for customizing the histogram
(for example, by adding a title).

Playing with breaks

R chooses the number of intervals it considers most useful to represent the
data, but you can disagree with what R does and choose the breaks yourself.
For this, you use the `breaks` argument of the `hist()` function.

You can specify the breaks in a couple of ways:

- ✔ **You can tell R the number of bars you want in the histogram by giving
  a single number as the argument.** Just keep in mind that R will still
  decide whether that's actually reasonable, and it tries to cut up the
  range using nice rounded numbers.

- ✔ **You can tell R exactly where to put the breaks by giving a vector with
  the break points as a value to the `breaks` argument.**

So, if you don't agree with R and you want to have bars representing the
intervals 5 to 15, 15 to 25, and 25 to 35, you can do this with the following
code:

```
> hist(cars$mpg, breaks = c(5, 15, 25, 35))
```

The resulting plot is on the right side of Figure 14-1.

You also can give the name of the algorithm R has to use to determine the number of breaks as the value for the `breaks` argument. You can find more information on those algorithms on the Help page `?hist`. Try to experiment with those algorithms a bit to check which one works the best.

Using frequencies or densities

By breaking up your data in intervals, you still lose some information, albeit a lot less than when just looking at the descriptives you calculate in the previous sections. Still, the most complete way of describing your data is by estimating the *probability density function* (PDF) or *density* of your variable.

If this concept is unfamiliar to you, don't worry. Just remember that the density is proportional to the chance that any value in your data is approximately equal to that value. In fact, for a histogram, the density is calculated from the counts, so the only difference between a histogram with frequencies and one with densities, is the scale of the *y*-axis. For the rest, they look exactly the same.

Creating a density plot

You can estimate the density function of a variable using the `density()` function. The output of this function itself doesn't tell you that much, but you can easily use it in a plot. For example, you can get the density of the mileage variable `mpg` like this:

```
> mpgdens <- density(cars$mpg)
```

The object you get this way is a list containing a lot of information you don't really need to look at. But that list makes plotting the density as easy as saying "plot the density":

```
> plot(mpgdens)
```

You see the result of this command on the left side of Figure 14-2. The plot looks a bit rough on the edges, but you can polish it with the tricks shown in Chapter 16. The important thing is to see how your data comes out. The density object is plotted as a line, with the actual values of your data on the *x*-axis and the density on the *y*-axis.

The `mpgdens` list object contains — among other things — a component called `x` and one called `y`. These represent the *x*- and *y*-coordinates for plotting the density. When R calculates the density, the `density()` function splits up your data in a large number of small intervals and calculates the density for the midpoint of each interval. Those midpoints are the values for `x`, and the calculated densities are the values for `y`.

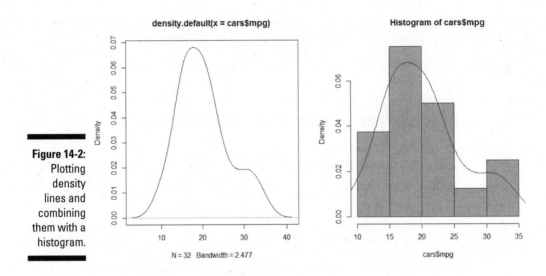

Plotting densities in a histogram

Remember that the `hist()` function returns the counts for each interval. Now the chance that a value lies within a certain interval is directly proportional to the counts. The more values you have within a certain interval, the greater the chance that any value you picked is lying in that interval.

So, instead of plotting the counts in the histogram, you could just as well plot the densities. R does all the calculations for you — the only thing you need to do is set the `freq` argument of `hist()` to `FALSE`, like this:

```
> hist(cars$mpg, col = "grey", freq = FALSE)
```

Now the plot will look exactly the same as before; only the values on the *y*-axis are different. The scale on the *y*-axis is set in such a way that you can add the density plot over the histogram. For that, you use the `lines()` function with the density object as the argument. So, you can, for example, fancy up the previous histogram a bit further by adding the estimated density using the following code immediately after the previous command:

```
> lines(mpgdens)
```

You see the result of these two commands on the right side of Figure 14-2. You get more information on how the `lines()` function works in Chapter 16. For now, just remember that `lines()` uses the `x` and `y` elements from the density object `mpgdens` to plot the line.

Describing Multiple Variables

Until now, you looked at a single variable from your dataset each time. All these statistics and plots tell part of the story, but when you have a dataset with multiple variables, there's a lot more of the story to be told. Taking a quick look at the summary of the complete dataset can warn you already if something went wrong with the data gathering and manipulation. But what statisticians really go after is the story told by the relation between the variables. And that story begins with describing these relations.

Summarizing a complete dataset

If you need a quick overview of your dataset, you can, of course, always use `str()` and look at the structure. But this tells you something only about the classes of your variables and the number of observations. Also, the function `head()` gives you, at best, an idea of the way the data is stored in the dataset.

Getting the output

To get a better idea of the distribution of your variables in the dataset, you can use the `summary()` function like this:

```
> summary(cars)
      mpg             cyl              am         gear
 Min.   :10.40   Min.   :4.000   auto  :13   3:15
 1st Qu.:15.43   1st Qu.:4.000   manual:19   4:12
 Median :19.20   Median :6.000               5: 5
 Mean   :20.09   Mean   :6.188
 3rd Qu.:22.80   3rd Qu.:8.000
 Max.   :33.90   Max.   :8.000
```

 The `summary()` function works best if you just use R interactively at the command line for scanning your dataset quickly. You shouldn't try to use it within a custom function you wrote yourself. In that case, you'd better use the functions from the first part of this chapter to get the desired statistics.

The output of `summary()` shows you for every variable a set of descriptive statistics, depending on the type of the variable:

- **Numerical variables:** the range, quartiles, median, and mean.
- **Factor variables:** a table with frequencies.
- **Numerical and factor variables:** the number of missing values, if there are any.
- **Character variables:** `summary()` doesn't give you any information at all apart from the length and the class (which is `'character'`).

Fixing a problem

Did you see the weird values for the variable `cyl`? A quick look at the summary can tell you there's something fishy going on, as, for example, the minimum and the first quartile have exactly the same value. In fact, the variable `cyl` has only three values and would be better off as a factor. So, let's put that variable out of its misery:

```
> cars$cyl <- as.factor(cars$cyl)
```

Now you can use it correctly in the remainder of this chapter.

Plotting quantiles for subgroups

Often you want to split up this analysis for different subgroups in order to compare them. You need to do this if you want to know how the average lip size compares between male and female kissing gouramis (great fish by the way!) or, in the case of our example, you want to know whether the number of cylinders in a car influences the mileage.

Of course you can use `tapply()` to calculate any of the descriptive statistics for subgroups defined by a factor variable. In Chapter 13, you do exactly that. But in R you find some more tools for summarizing descriptive statistics for different subgroups.

One way to quickly compare groups is to construct a box-and-whisker plot from the data. You could construct this plot by calculating the range, the quartiles, and the median for each group, but luckily you can just tell R to do all that for you. For example, if you want to know how the mileage compares between cars with a different number of cylinders, you simply use the `boxplot()` function to get the result shown in Figure 14-3:

```
> boxplot(mpg ~ cyl, data = cars)
```

You supply a simple formula as the first argument to `boxplot()`. This formula reads as "plot boxes for the variable `mpg` for the groups defined by the variable `cyl`." You find more information on the formula interface for functions in Chapter 13.

This plot uses quantiles to give you an idea of how the data is spread within each subgroup. The line in the middle of each box represents the median, and the edges of the box represent the first and the third quartiles. The whiskers extend to either the minimum and the maximum of the data or 1.5 times the distance between the first and the third quartiles, whichever is smaller.

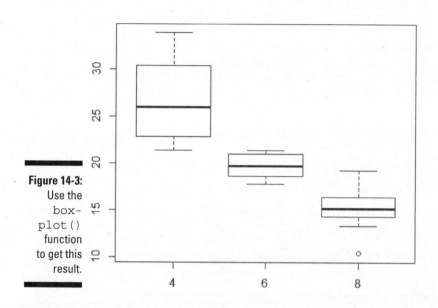

Figure 14-3:
Use the
box-
plot()
function
to get this
result.

To be completely correct, the edges of the box represent the lower
and upper hinges from the five-number summary, calculated using the
fivenum() function. They're equal to the quartiles only if you have an odd
number of observations in your data. Otherwise, the results of fivenum()
and quantile() may differ a bit due to differences in the details of the
calculation.

You can let the whiskers always extend to the minimum and the maximum by
setting the range argument of the boxplot() function to 0.

Extracting the data from the plots

The hist() and boxplot() functions have another incredibly nice feature: You can get access
to all the information R uses to plot the histogram or box plot and use it in further calculations.
Getting that information is as easy as assigning the output of the function to an object. For example,
you get the information on the breaks, counts, and density in a histogram like this:

```
> mpghist <- hist(cars$mpg)
```

This still plots your histogram, and in addition you create an object that contains a list with — among
other things — the components breaks, counts, and density. For a box plot, you can do exactly the
same and get an object that contains a list with — among other things — the components stats
and n, representing the used statistics and the number of cases in each category. On the Help

(continued)

(continued)

pages for `hist()` and `boxplot()`, you find more information on the list components in the "Value" sections. To avoid the plot being created, you can set the argument plot to `FALSE` in either function.

All that information you could, of course, also get using other functions in R. It can help, though, to quickly add some extra information to a plot. For example, you can add the number of cases for each box to a box plot like this:

```
> mpgbox <- boxplot(mpg ~ cyl, data = cars)
> n <- nlevels(as.factor(cars$cyl))
> text(1:n, mpgbox$stats[1, ], paste("n =", mpgbox$n), pos = 1)
```

With this code, you add a text value under the lower whisker. The *x*-coordinates 1 through *n* coincide with the middle of each box. You get the *y*-coordinates from the `stats` element in the `mpgbox` object, which tells you where the lower whisker is. The argument `pos=1` in the `text()` function places the text under the coordinates. You can try playing around with it yourself. While you're at it, check Chapter 16 for some more tips on manipulating plots.

Tracking correlations

Statisticians love it when they can link one variable to another. Sunlight, for example, is detrimental to trouser length: The more the sun shines, the shorter the trouser length — in summer people wear shorts; in winter they wear long trousers. We say that the number of hours of sunshine correlates with trouser length. Obviously, there isn't really a direct causal relationship here — you won't find shorts during the summer in polar regions. But, in many cases, the search for causal relationships starts with looking at correlations.

To illustrate, take a look at the famous `iris` dataset in R. One of the founding fathers of statistics, Sir Ronald Fisher, used this dataset to illustrate how multiple measurements can be used to discriminate between different species. This dataset contains five variables, as you can see by using the `names()` function:

```
> names(iris)
[1] "Sepal.Length" "Sepal.Width"  "Petal.Length"
[4] "Petal.Width"  "Species"
```

It contains measurements of flower characteristics for three species of iris and from 50 flowers for each species. Two variables describe the sepals (`Sepal.Length` and `Sepal.Width`), two other variables describe the petals (`Petal.Length` and `Petal.Width`), and the last variable (`Species`) is a factor indicating from which species each flower comes.

Looking at relations

Although looks can be deceiving, you want to eyeball your data before digging deeper into it. In Chapter 16, you create scatterplots for two variables. To plot a grid of scatterplots for all combinations of two variables in your dataset, you can simply use the `plot()` function on your data frame, like this:

```
> plot(iris[-5])
```

Because scatterplots are useful only for continuous variables, you can drop all variables that are not continuous. Too many variables in the plot matrix make the plots difficult to see. In the previous code, you drop the variable `Species`, because that's a factor.

You can see the result of this simple line of code in Figure 14-4. The variable names appear in the squares on the diagonal, indicating which variables are plotted along the *x*-axis and the *y*-axis. For example, the second plot on the third row has `Sepal.Width` on the *x*-axis and `Petal.Length` on the *y*-axis.

When the `plot()` function notices that you pass a data frame as an argument, it calls the `pairs()` function to create the plot matrix. This function offers you a lot more flexibility. For example, on the Help page `?pairs`, you find some code that adds a histogram on the diagonal plots. Check out the examples on the Help page for some more tricks.

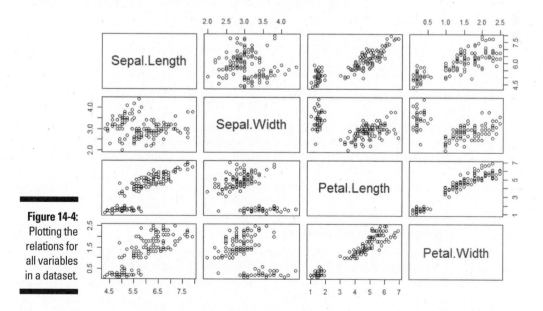

Figure 14-4: Plotting the relations for all variables in a dataset.

Getting the numbers

The amount in which two variables vary together can be described by the *correlation coefficient*. You get the correlations between a set of variables in R very easily by using the `cor()` function. You simply add the two variables you want to examine as the arguments. For example, if you want to check how much the petal width correlates with the petal length, try:

```
> with(iris, cor(Petal.Width, Petal.Length))
[1] 0.9628654
```

This tells you that the relation between the petal width and the petal length is almost a perfect line, as you also can see in the fourth plot of the third row in Figure 14-4.

Calculating correlations for multiple variables

You also can calculate the correlation among multiple variables at once, much in the same way as you can plot the relations among multiple variables. So, for example, you can calculate the correlations that correspond with the plot in Figure 14-4 with the following line:

```
> iris.cor <- cor(iris[-5])
```

As always, you can save the outcome of this function in an object. This lets you examine the structure of the function output so you can figure out how you can use it in the rest of your code. Here's a look at the structure of the object `iris.cor`:

```
> str(iris.cor)
 num [1:4, 1:4] 1 -0.118 0.872 0.818 -0.118 . . .
 - attr(*, "dimnames")=List of 2
  ..$ : chr [1:4] "Sepal.Length" "Sepal.Width" "Petal.Length" "Petal.Width"
  ..$ : chr [1:4] "Sepal.Length" "Sepal.Width" "Petal.Length" "Petal.Width"
```

This output tells you that `iris.cor` is a matrix with the names of the variables as both row names and column names. To find the correlation of two variables in that matrix, you can use the names as indices — for example:

```
> iris.cor["Petal.Width", "Petal.Length"]
[1] 0.9628654
```

Dealing with missing values

The `cor()` function can deal with missing values in multiple ways. For that, you set the argument `use` to one of the possible text values. The value for the use argument is especially important if you calculate the correlations of the variables in a data frame. By setting this argument to different values, you can

- ✔ **Use all observations by setting** use="everything". This means that if any variable contains any NA values, the resulting correlation is also NA. This is the default.

- ✔ **Exclude all observations that have NA for at least one variable.** For this, you set use="complete.obs". Note that this may leave you with only a few observations if missing values are spread through the complete dataset.

- ✔ **Exclude observations with NA values for every pair of variables you examine.** For that, you set the argument use="pairwise". This ensures that you can calculate the correlation for every pair of variables without losing information because of missing values in the other variables.

You might think that use = "pairwise" should be the default choice, because it uses the most data. However, this could lead to mathematical problems downstream. (For the mathematical whiz kids: The resulting correlation matrix is not guaranteed to be *positive definite*).

In fact, you can calculate different measures of correlation. By default, R calculates the standard Pearson correlation coefficient. For data that is not normally distributed, you can use the cor() function to calculate the Spearman rank correlation, or Kendall's tau. For this, you have to set the method argument to the appropriate value. You can find more information about calculating the different correlation statistics on the Help page ?cor. For more formal testing of correlations, look at the cor.test() function and the related Help page.

Working with Tables

In the "Describing Categories" section, earlier in this chapter, you use tables to summarize one categorical variable. But tables can easily describe more variables at once. You may want to know how many men and women teach in each department of your university (although that's not the most traditional criterion for choosing your major).

Creating a two-way table

A *two-way table* is a table that describes two categorical variables together. It contains the number of cases for each combination of the categories in both variables. The analysis of categorical data always starts with tables, and R gives you a whole toolset to work with them. But first, you have to create the tables.

Creating a table from two variables

For example, you want to analyze the number of gears (3, 4, or 5) with gearbox type (automatic or manual). You can do this again using the `table()` function with two arguments, like this:

```
> with(cars, table(am, gear))

            3   4   5
    auto    0   8   5
    manual 15   4   0
```

The levels of the variable you give as the first argument are the row names, and the levels of the variable you give as the second argument are the column names. In the table, you get the counts for every combination. For example, you can count 15 cars with manual gearboxes and three gears.

Creating tables from a matrix

Researchers also use tables for more serious business, like finding out whether a certain behavior (like smoking) has an impact on the risk of getting an illness (for example, lung cancer). This way you have four possible cases: risk behavior and sick, risk behavior and healthy, no risk behavior and healthy, or no risk behavior and sick.

Often the result of such a study consists of the counts for every combination. If you have the counts for every case, you can very easily create the table yourself, like this:

```
> trial <- matrix(c(34, 11, 9, 32), ncol = 2)
> colnames(trial) <- c("sick", "healthy")
> rownames(trial) <- c("risk", "no_risk")
> trial.table <- as.table(trial)
```

With this code, you do the following:

1. **Create a matrix with the number of cases for every combination of sick/healthy and risk/no risk behavior.**

2. **Add column names to point out which category the counts are for.**

3. **Convert that matrix to a table.**

The result looks like this:

```
> trial.table
        sick healthy
risk      34       9
no_risk   11      32
```

A table like `trial.table` can be seen as a summary of two variables. One variable indicates if the person is sick or healthy, and the other variable indicates whether the person shows risky behavior.

Extracting the numbers

Although tables and matrices are two different beasts, you can treat a two-way table like a matrix in most situations. This becomes handy if you want to extract values from the table. If you want to know how many people were sick and showed risk behavior, you simply do the following:

```
> trial.table["risk", "sick"]
[1] 34
```

All the tricks with indices that we cover in Chapters 4 and 7 work on tables, too. A table of a single variable reacts the same as a vector, and a two-way table reacts the same as a matrix.

Converting tables to a data frame

The resulting object `trial.table` *looks* exactly the same as the matrix `trial`, but it really isn't. The difference becomes clear when you transform these objects to a data frame. Take a look at the outcome of this code:

```
> trial.df <- as.data.frame(trial)
> str(trial.df)
'data.frame': 2 obs. of  2 variables:
 $ sick   : num  34 11
 $ healthy: num  9 32
```

Here you get a data frame with two variables (`sick` and `healthy`) with each two observations. On the other hand, if you convert the table to a data frame, you get the following result:

```
> trial.table.df <- as.data.frame(trial.table)
> str(trial.table.df)
'data.frame': 4 obs. of  3 variables:
 $ Var1: Factor w/ 2 levels "risk","no_risk": 1 2 1 2
 $ Var2: Factor w/ 2 levels "sick","healthy": 1 1 2 2
 $ Freq: num  34 11 9 32
```

The `as.data.frame()` function converts a table to a data frame in a format that you need for regression analysis on count data. If you need to summarize the counts first, you use `table()` to create the desired table.

Now you get a data frame with three variables. The first two — `Var1` and `Var2` — are factor variables for which the levels are the values of the rows and the columns of the table, respectively. The third variable — `Freq` — contains the frequencies for every combination of the levels in the first two variables.

In fact, you also can create tables in more than two dimensions by adding more variables as arguments, or by transforming a multidimensional array to a table using `as.table()`. You can access the numbers the same way you do for multidimensional arrays, and the `as.data.frame()` function creates as many factor variables as there are dimensions.

Looking at margins and proportions

In categorical data analysis, many techniques use the *marginal totals* of the table in the calculations. The marginal totals are the total counts of the cases over the categories of interest. For example, the marginal totals for behavior would be the sum over the rows of the table `trial.table`.

Adding margins to the table

R allows you to extend a table with the marginal totals of the rows and columns in one simple command. Use the `addmargins()` function, like this:

```
> addmargins(trial.table)
        sick healthy Sum
risk      34       9  43
no_risk   11      32  43
Sum       45      41  86
```

You also can add the margins for only one dimension by specifying the `margin` argument for the `addmargins()` function. For example, to get only the marginal counts for the behavior, you do the following:

```
> addmargins(trial.table, margin = 2)
        sick healthy Sum
risk      34       9  43
no_risk   11      32  43
```

The `margin` argument takes a number or a vector of numbers, but it can be a bit confusing. The margins are numbered the same way as in the `apply()` function. So 1 stands for rows and 2 for columns. To add the column margin, you need to set `margin` to 2, but this column margin contains the row totals.

Calculating proportions

You can convert a table with counts to a table with proportions very easily using the `prop.table()` function. This also works for multi-way tables. If you want to know the proportions of observations in every cell of the table to the total number of cases, you simply do the following:

```
> prop.table(trial.table)
            sick    healthy
risk    0.3953488 0.1046512
no_risk 0.1279070 0.3720930
```

This tells you that, for example, 10.4 percent of the people in the study were healthy, even when they showed risk behavior.

Calculating proportions over columns and rows

But what if you want to know which fraction of people with risk behavior got sick? Then you don't have to calculate the proportions by dividing the counts by the total number of cases for the whole dataset; instead, you divide the counts by the marginal totals.

R lets you do this very easily using, again, the `prop.table()` function, but this time specifying the `margin` argument.

Take a look at the table again. You want to calculate the proportions over each row, because each row represents one category of behavior. So, to get the correct proportions, specify `margin=1` like this:

```
> prop.table(trial.table, margin = 1)
            sick    healthy
risk    0.7906977 0.2093023
no_risk 0.2558140 0.7441860
```

In every row, the proportions sum to 1. Notice that 79 percent of the people showing risk behavior got sick. Well, it isn't big news that risky behavior can cause diseases, and the proportions shown in the last result point in that direction. Yet, scientists believe you only if you can back it up in a more objective way. That's the point at which you should consider doing some statistical testing. We show you how in Chapter 15.

Chapter 15

Testing Differences and Relations

*I*t's one thing to describe your data and plot a few graphs, but if you want to draw conclusions from these graphs, people expect a bit more proof. This is where the data analysts chime in and start pouring p-values generously over reports and papers. These p-values summarize the conclusions of statistical tests, basically indicating how likely it is that the result you see is purely due to chance. The story is a bit more complex — but for that you need to take a look at a statistics handbook like *Statistics For Dummies,* 2nd Edition, by Deborah J. Rumsey, PhD (Wiley).

R really shines when you need some serious statistical number crunching. Statistics is the alpha and omega of this language, but why have we waited until Chapter 15 to cover some of that? There are two very good reasons why we wait to talk about statistics until now:

✔ You can start with the statistics only after you've shaped your data into the right format, so you need to get that down first.

✔ R contains an overwhelming amount of advanced statistical techniques, many of which come with their own books and manuals.

Luckily, many packages follow the same principles regarding the user interface. So, instead of trying to cover all of what's possible, in this chapter we introduce you to some basic statistical tests and explain the interfaces in more detail so you get familiar with that.

Taking a Closer Look at Distributions

The *normal distribution* (also known as the *Gaussian distribution* or *bell curve*) is a key concept in statistics. Much statistical inference is based on the assumption that, at some point in your calculations, you have values that are distributed normally. Testing whether the distribution of your data follows this bell curve closely enough is often one of the first things you do before you choose a test to test your hypothesis.

If you're not familiar with the normal distribution, check out *Statistics For Dummies,* 2nd Edition, by Deborah J. Rumsey, PhD (Wiley), which devotes a whole chapter to this concept.

Observing beavers

The biologist and statistician Penny Reynolds observed some beavers for a complete day and measured their body temperature every ten minutes. She also wrote down whether the beavers were active at that moment. You find the measurements in the datasets `beaver1` and `beaver2`. (See ?beavers for more information.) In the following examples, we use `beaver2`:

```
> str(beaver2)
'data.frame': 100 obs. of  4 variables:
 $ day  : num  307 307 307 307 307 . . .
 $ time : num  930 940 950 1000 1010 . . .
 $ temp : num  36.6 36.7 . . .
 $ activ: num  0 0 0 0 0 . . .
```

If you want to know whether there's a difference between the average body temperature during periods of activity and periods without, you first have to make sure all variables are of a suitable type. In this case, the variable `activ` is a numeric variable but actually describes categories. So before doing anything else, you transform that variable to a factor like this:

```
> transform.beaver <- transform(beaver2,
+     activ = factor(activ, labels = c("no", "yes"))
+ )
```

Then you choose a test. To know which test is appropriate, you need to find out if the temperature is distributed normally during both periods. So, let's take a closer look at the distributions.

Testing normality graphically

You could, of course, plot a histogram for every sample you want to look at. You can use the histogram() function pretty easily to plot histograms for different groups. (This function is part of the lattice package you use in Chapter 17.)

Using the formula interface, you can plot two histograms in Figure 15-1 at once using the following code:

```
> library("lattice")
> histogram(~temp | activ, data = transform.beaver)
```

You find more information about the formula interface in Chapter 13, but let's go over this formula once more. The histogram() function uses a one-sided formula, so you don't specify anything at the left side of the tilde (~). On the right side, you specify:

✔ **Which variable the histogram should be created for:** In this case, that's the variable temp, containing the body temperature.

✔ **After the vertical line (|),the factor by which the data should be split:** In this case, that's the variable activ that has a value yes if the beaver was active and no if it was not.

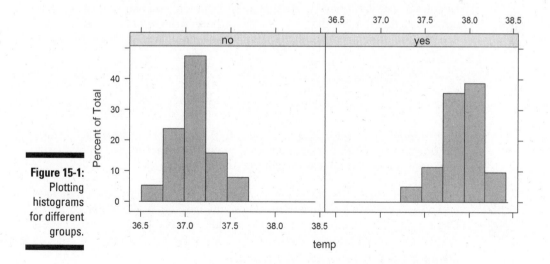

Figure 15-1:
Plotting
histograms
for different
groups.

You can read the vertical line (|) in the formula interface as "conditional on." It's also used in that context in the formula interfaces of more advanced statistical functions.

Using quantile plots

Still, histograms leave much to the interpretation of the viewer. A better graphical way to tell whether your data is distributed normally is to look at a so-called quantile-quantile (QQ) plot.

With this technique, you plot quantiles against each other. If you compare two samples, for example, you simply compare the quantiles of both samples. Or, to put it a bit differently, R does the following to construct a QQ plot:

✔ It sorts the data of both samples.

✔ It plots these sorted values against each other.

If both samples don't contain the same number of values, R calculates extra values by interpolation for the smallest sample to create two samples of the same size.

Comparing two samples

Of course, you don't have to do that all by yourself, you can simply use the qqplot() function for that. So, to check whether the temperatures during activity and during rest are distributed equally, you simply do the following:

```
> with(transform.beaver,
+       qqplot(temp[activ == "yes"],
+              temp[activ == "no"])
+ )
```

This creates a plot where the ordered values are plotted against each other, as shown in Figure 15-2. You can use all the tricks from Chapter 16 to change axis titles, color and appearance of the points, and so on.

Between the square brackets, you can use a logical vector to select the cases you want. Here you select all cases where the variable activ equals 1 for the first sample, and all cases where that variable equals 0 for the second sample.

Using a QQ plot to check for normality

In most cases, you don't want to compare two samples with each other, but compare a sample with a theoretical sample that comes from a certain distribution (for example, the normal distribution).

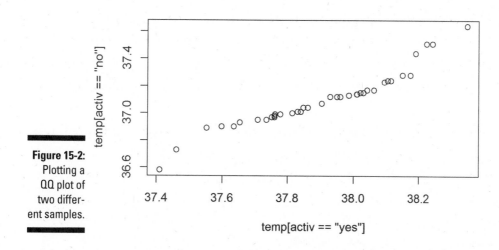

Figure 15-2:
Plotting a
QQ plot of
two differ-
ent samples.

To make a QQ plot this way, R has the special qqnorm() function. As the name implies, this function plots your sample against a normal distribution. You simply give the sample you want to plot as a first argument and add any of the graphical parameters from Chapter 16 you like. R then creates a sample with values coming from the *standard* normal distribution, or a normal distribution with a mean of zero and a standard deviation of one. With this second sample, R creates the QQ plot as explained before.

R also has a qqline() function, which adds a line to your normal QQ plot. This line makes it a lot easier to evaluate whether you see a clear deviation from normality. The closer all points lie to the line, the closer the distribution of your sample comes to the normal distribution. The qqline() function also takes the sample as an argument.

Now you want to do this for the temperatures during both the active and the inactive period of the beaver. You can use the qqnorm() function twice to create both plots. For the inactive periods, you can use the following code:

```
> with(transform.beaver, {
+    qqnorm(temp[activ == "no"], main = "Inactive")
+    qqline(temp[activ == "no"])
+ })
```

You can do the same for the active period by changing the value "no" to "yes". The resulting plots you see in Figure 15-3.

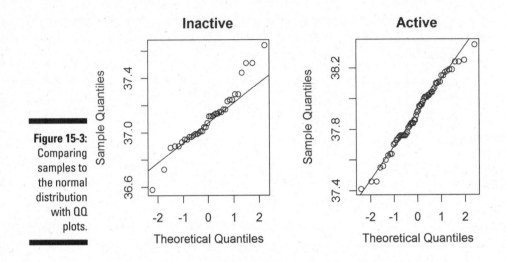

Figure 15-3:
Comparing
samples to
the normal
distribution
with QQ
plots.

Testing normality in a formal way

Graphical methods for checking normality leave much to your own interpretation. If you show any of these plots to ten different statisticians, you can end up with ten different opinions. That's quite an achievement when you expect a simple yes or no, but statisticians don't do simple answers.

On the contrary, everything in statistics revolves around measuring uncertainty. This uncertainty is often summarized in a probability — often called a *p-value* — and to calculate this probability, you need a formal test.

Probably the most widely used test for normality is the Shapiro-Wilks test. The function to perform this test, conveniently called shapiro.test(), couldn't be easier to use. You give the sample as the one and only argument, as in the following example:

```
> shapiro.test(transform.beaver$temp)

 Shapiro-Wilks normality test

data:  transform.beaver$temp
W = 0.9334, p-value = 7.764e-05
```

This function returns a list object, and the p-value is contained in an component called p.value. So, for example, you can extract the p-value simply by using the following code:

```
> result <- shapiro.test(transform.beaver$temp)
> result$p.value
[1] 7.763782e-05
```

This p-value tells you what the chances are that the sample comes from a normal distribution. The lower this value, the smaller the chance. Statisticians typically use a value of 0.05 as a cutoff, so when the p-value is lower than 0.05, you can conclude that the sample deviates from normality. In the preceding example, the p-value is clearly lower than 0.05 — and that shouldn't come as a surprise; the distribution of the temperature shows two separate peaks (refer to Figure 15-1). This is nothing like the bell curve of a normal distribution.

When you choose a test, you may be more interested in the normality in each sample. You can test both samples in one line using the `tapply()` function, like this:

```
> with(transform.beaver, tapply(temp, activ, shapiro.test))
```

This code returns the results of a Shapiro-Wilks test on the temperature for every group specified by the variable `activ`.

Statisticians sometimes refer to the Kolmogorov-Smirnov test for testing normality. You carry out the test by using the `ks.test()` function in base R. But this R function is not suited to test deviation from normality; you can use it only to compare different distributions.

Comparing Two Samples

Comparing groups is one of the most basic problems in statistics. If you want to know if extra vitamins in the diet of cows is increasing their milk production, you give the normal diet to a control group and extra vitamins to a test group, and then you compare the milk production in two groups. By comparing the mileage between cars with automatic gearboxes and those with manual gearboxes, you can find out which one is the more economical option.

Testing differences

R gives you two standard tests for comparing two groups with numerical data: the *t-test* with the `t.test()` function, and the *Wilcoxon test* with the `wilcox.test()` function. If you want to use the `t.test()` function, you first have to check, among other things, whether both samples are normally distributed using any of the methods from the previous section. For the Wilcoxon test, this isn't necessary.

Carrying out a t-test

Let's take another look at the data of that beaver. If you want to know if the average temperature differs between the periods the beaver is active and inactive, you can do so with a simple command:

```
> t.test(temp ~ activ, data = transform.beaver)

              Welch Two-Sample t-test

data:  temp by activ
t = -18.5479, df = 80.852, p-value < 2.2e-16
alternative hypothesis: true difference in means is not equal to 0
95 percent confidence interval:
 -0.8927106 -0.7197342
sample estimates:
mean in group 0 mean in group 1
      37.09684        37.90306
```

Normally, you can only carry out a t-test on samples for which the variances are approximately equal. R uses Welch's variation on the t-test, which corrects for unequal variances.

You get a whole lot of information here:

- ✔ The second line gives you the test statistic (t for this test), the degrees of freedom (df), and the corresponding p-value. The very small p-value indicates that the means of both samples differ significantly.

- ✔ The alternative hypothesis tells you what you can conclude if the p-value is lower than the limit for significance. Generally, scientists consider the alternative hypothesis to be true if the p-value is lower than 0.05.

- ✔ The 95 percent confidence interval is the interval that contains the difference between the means with 95 percent probability, so in this case the difference between the means lies probably between 0.72 and 0.89.

- ✔ The last line gives you the means of both samples.

You read the formula `temp ~ activ` as "evaluate temp within groups determined by `activ`." Alternatively, you can use two separate vectors for the samples you want to compare and pass both to the function, as in the following example:

```
> with(transform.beaver,
+      t.test(temp[activ == "yes"],
+             temp[activ == "no"]))
```

Dropping assumptions

In some cases, your data deviates significantly from normality and you can't use the `t.test()` function. For those cases, you have the `wilcox.test()` function, which you use in exactly the same way, as shown in the following example:

```
> wilcox.test(temp ~ activ, data = transform.beaver)
```

This gives you the following output:

```
  Wilcoxon rank-sum test with continuity correction

data:  temp by activ
W = 15, p-value < 2.2e-16
alternative hypothesis: true location shift is not equal to 0
```

Again, you get the value for the test statistic (`W` in this test) and a p-value. Under that information, you read the alternative hypothesis, and that differs a bit from the alternative hypothesis of a t-test. The Wilcoxon test looks at whether the center of your data (the location) differs between both samples.

With this code, you perform the Wilcoxon rank-sum test or Mann-Whitney U test. Both tests are completely equivalent, so R doesn't contain a separate function for the Mann-Whitney U test.

Testing direction

In both previous examples, you test whether the samples differ without specifying in which way. Statisticians call this a *two-sided test.* Imagine you don't want to know whether body temperature differs between active and inactive periods, but whether body temperature is lower during inactive periods.

To do this, you have to specify the argument alternative in either the `t.test()` or `wilcox.test()` function. This argument can take three values:

- ✔ By default, it has the value `"two.sided"`, which means you want the standard two-sided test.

- ✔ If you want to test whether the mean (or location) of the first group is lower, you give it the value `"less"`.

- ✔ If you want to test whether that mean is bigger, you specify the value `"greater"`.

If you use the formula interface for these tests, the groups are ordered in the same order as the levels of the factor you use. You have to take that into account to know which group is seen as the first group. If you give the data for both groups as separate vectors, the first vector is the first group.

Comparing paired data

When testing differences between two groups, you can have either paired or unpaired data. Paired data comes from experiments where two different treatments were given to the same subjects.

For example, researchers give ten people two variants of a sleep medicine. Each time the researchers record the difference in hours of sleep with and without the drugs. Because each person receives both variants, the data is paired. You find the data of this experiment in the dataset sleep, which has three variables:

- ✔ A numeric variable extra, which gives the extra hours of sleep after the medication is taken
- ✔ A factor variable group that tells which variant the person took
- ✔ A factor variable id that indicates the ten different test persons

Now they want to know whether both variants have a different effect on the length of the sleep. Both the t.test() and the wilcox.test() functions have an argument paired that you can set to TRUE in order to carry out a test on paired data. You can test differences between both variants using the following code:

```
> t.test(extra ~ group, data = sleep, paired = TRUE)
```

This gives you the following output:

```
  Paired t-test

data:  extra by group
t = -4.0621, df = 9, p-value = 0.002833
alternative hypothesis: true difference in means is not equal to 0
95 percent confidence interval:
 -2.4598858 -0.7001142
sample estimates:
mean of the differences
               -1.58
```

Unlike the unpaired test, you don't get the means of both groups; instead, you get a single mean of the differences.

Testing Counts and Proportions

Many research questions revolve around counts instead of continuous numerical measurements. Counts can be summarized in tables and subsequently analyzed. In the following section, you use some of the basic tests for counts and proportions contained in R. Be aware, though, that this is just the tip of the iceberg; R has a staggering amount of statistical procedures for categorical data available.

If you need more background on the statistics, check the book *Categorical Data Analysis*, 3rd Edition, by Alan Agresti (Wiley-Interscience). You find more information, including links to datasets and a pdf file with R code for the examples in the book, on the book's website at http://www.stat.ufl.edu/~aa/cda/cda.html. The related book *An Introduction to Categorical Data Analysis*, also by Alan Agresti (Wiley) offers a more entry-level introduction to the topic.

Checking out proportions

Let's look at an example to illustrate the basic tests for proportions.

The following example is based on real research, published by Robert Rutledge, MD, and his colleagues in the *Annals of Surgery* (1993).

In a hospital in North Carolina, the doctors registered the patients who were involved in a car accident and whether they used seat belts. The following matrix represents the number of survivors and deceased patients in each group:

```
> survivors <- matrix(c(1781, 1443, 135, 47), ncol = 2)
> colnames(survivors) <- c("survived", "died")
> rownames(survivors) <- c("no seat belt", "seat belt")
> survivors
             survived died
no seat belt     1781  135
seat belt        1443   47
```

To know whether seat belts made a difference in the chances of surviving, you can carry out a proportion test. This test calculates how probable it is that both proportions are the same. A low p-value tells you that both proportions probably differ from each other. To test this in R, you can use the prop.test() function on the preceding matrix:

```
> result.prop <- prop.test(survivors)
```

You also can use the `prop.test()` function on tables or vectors. If you use it with vectors, remember that the first vector has to be the number of successes, and the second number has to be the *total* number of cases.

The `prop.test()` function then gives you the following output:

```
> result.prop

  2-sample test for equality of proportions with continuity correction

data:  survivors
X-squared = 24.3328, df = 1, p-value = 8.105e-07
alternative hypothesis: two.sided
95 percent confidence interval:
 -0.05400606 -0.02382527
sample estimates:
   prop 1    prop 2
0.9295407 0.9684564
```

This test report is almost identical to the one from `t.test()` and contains essentially the same information. At the bottom, R prints for you the proportion of people who survived in each group. The p-value tells you how likely it is that both the proportions are equal. So, you see that the chance of dying in a hospital after a crash is lower if you're wearing a seat belt at the time of the crash. R also reports the confidence interval of the difference between the proportions.

Analyzing tables

You can use the `prop.test()` function for matrices and tables. For `prop.test()`, these tables need to have two columns with the number of counts for the two possible outcomes like the matrix `survivors` from the previous section.

Testing contingency of tables

Alternatively, you can use the `chisq.test()` function to analyze tables with a chi-squared (χ^2) contingency test. To do this on the matrix with the seat-belt data, you simply do the following:

```
> chisq.test(survivors)
```

This returns the following output:

```
  Pearson's Chi-squared test with Yates' continuity correction

data:  survivors
X-squared = 24.3328, df = 1, p-value = 8.105e-07
```

The values for the statistic (X-squared), the degrees of freedom, and the p-value are exactly the same as with the prop.test() function. That's to be expected, because — in this case, at least — both tests are equivalent.

Testing tables with more than two columns

Unlike the prop.test() function, the chisq.test() function can deal with tables with more than two columns. To illustrate this, let's take a look at the table HairEyeColor. You can see its structure with the following code:

```
> str(HairEyeColor)
 table [1:4, 1:4, 1:2] 32 53 10 3 11 50 10 30 10 25 . . .
 - attr(*, "dimnames")=List of 3
  ..$ Hair: chr [1:4] "Black" "Brown" "Red" "Blond"
  ..$ Eye : chr [1:4] "Brown" "Blue" "Hazel" "Green"
  ..$ Sex : chr [1:2] "Male" "Female"
```

So, the table HairEyeColor has three dimensions: one for hair color, one for eye color, and one for sex. The table represents the distribution of these three features among 592 students.

The dimension names of a table are stored in an attribute called dimnames. As you can see from the output of the str() function, this is actually a list with the names for the rows/columns in each dimension. If this list is a named list, the names are used to label the dimensions. You can use the dimnames() function to extract or change the dimension names. (Go to the Help page ?dimnames for more examples.)

To check whether hair color and eye color are related, you can collapse the table over the first two dimensions using the margin.table() function to summarize hair and eye color for both genders. This function sums the values in some dimensions to give you a summary table with fewer dimensions. For that, you have to specify which margins you want to keep.

So, to get the table of hair and eye color, you use the following:

```
> HairEyeMargin <- margin.table(HairEyeColor, margin = c(1, 2))
> HairEyeMargin
       Eye
Hair    Brown Blue Hazel Green
  Black    68   20    15     5
  Brown   119   84    54    29
  Red      26   17    14    14
  Blond     7   94    10    16
```

Now you can simply check whether hair and eye color are related by testing it on this table:

```
> chisq.test(HairEyeMargin)

              Pearson's Chi-squared test

data:  HairEyeMargin
X-squared = 138.2898, df = 9, p-value < 2.2e-16
```

As expected, the output of this test tells you that some combinations of hair and eye color are more common than others. Not a big surprise, but you can use these techniques on other, more interesting research questions.

Extracting test results

Many tests in this chapter return a `htest` object. That type of object is basically a list with all the information about the test that has been carried out. All these `htest` objects contain at least a component `statistic` with the value of the statistic and a component `p.value` with the value of the p-value. You can see this easily if you look at the structure of the returned object. The object returned by `shapiro.test()` in the previous section looks like this:

```
> str(result)
List of 4
 $ statistic: Named num 0.933
  ..- attr(*, "names")= chr "W"
 $ p.value  : num 7.76e-05
 $ method   : chr "Shapiro-Wilk normality test"
 $ data.name: chr "transform.beaver$temp"
 - attr(*, "class")= chr "htest"
```

Because this `htest` objects are lists, you can use any of the list subsetting methods to extract the information. The following code, for example, extracts the p-value from the t-test on the beaver data:

```
> t.test(temp ~ activ, data = transform.beaver)$p.value
[1] 7.269112e-31
```

The extraction of information from the `htest` object also works with the results of many more `.test` functions, including the ones discussed in this chapter. You can check what kind of object a test returns by looking at the Help page for the test you want to use.

Working with Models

The tests we describe in the previous sections are all basic statistical tests. However, these days, much of statistics involves working with complex models. Base R already contains an extensive set of modeling tools, allowing you to do anything from simple linear models and analysis of variance to mixed models and time-series analysis. The nice thing about modeling functions in R is that they often work in a similar way.

Because this isn't a statistics book, we can't cover the details about assumption testing and model evaluation. If you aren't familiar with these concepts or the applied models, be sure to consult a decent source of information, or you'll run the risk of basing important decisions on inadequate models. The book *Applied Linear Statistical Models,* 5th Edition, by Michael Kutner et al (McGraw-Hill/Irwin), is very extensive but gives a good and thorough theoretical introduction on assumption testing and model evaluation.

In this section, we cover some basic models, and show you how to extract useful information from the resulting model objects.

Analyzing variances

An analysis of variance (ANOVA) is a very common technique used to compare the means between different groups. To illustrate this, take a look at the dataset `InsectSpray`:

```
> str(InsectSprays)
'data.frame': 72 obs. of  2 variables:
 $ count: num  10 7 20 14 14 12 10 23 17 20 . . .
 $ spray: Factor w/ 6 levels "A","B","C","D",..: 1 1 1 1 1 1 1 1 1 1 . . .
```

This dataset contains the results of an agricultural experiment. Six insecticides were tested on 12 fields each, and the researchers counted the number of pesky bugs that remained on each field. Now the farmers need to know if the insecticides make any difference, and if so, which one they should use. You answer this question by using the `aov()` function to perform an ANOVA.

Building the model

For this simple example, building the model is a piece of cake. You essentially want to model the means for the variable `count` as a function of the variable `spray`. You translate that to R like this:

```
> AOVModel <- aov(count ~ spray, data = InsectSprays)
```

You pass two arguments to the aov() function in this line of code:

- ✔ The formula count ~ spray, which reads as "count as a function of spray"
- ✔ The argument data, where you specify the data frame in which the variables in the formula can be found

Every modeling function returns a model object with a lot of information about the fitted model. Always put this model object in a variable. This way you don't have to refit the model when you need to perform extra calculations.

Looking at the object

As with every object, you can look at a model object just by typing its name in the console. If you do that for the object Model that you created in the preceding section, you see the following output:

```
> AOVModel
Call:
    aov(formula = count ~ spray, data = InsectSprays)

Terms:
                    spray Residuals
Sum of Squares   2668.833  1015.167
Deg. of Freedom        5        66

Residual standard error: 3.921902
Estimated effects may be unbalanced
```

This doesn't tell you that much, apart from the command (or the *call*) you used to build the model and some basic information on the fitting result.

In the output, you also read that the estimated effects may be unbalanced. This isn't a warning as described in Chapter 10 — it's a message that's built in by the author of the aov() function. This one can pop up in two situations:

- ✔ You don't have the same number of cases in every group.
- ✔ You didn't set orthogonal contrasts.

In this case, it's the second reason. You can continue with this model as we do now (that's also how those models are fitted in SPSS and SAS by default), or you can read the nearby sidebar, "Setting the contrasts," and use contrasts as the statistical experts who wrote R think you should.

Setting the contrasts

Before you can use the `aov()` function, you'd better set the *contrasts* you're going to use. Contrasts are very often forgotten about when doing ANOVA, but they generally help with interpreting the model and increase the accuracy of `aov()` and the helper functions.

What are those contrasts then? Factors are translated to a set of variables, with one fewer variable than the number of levels of the factor. Say you have a factor with three levels. R creates two variables, and each level of the factor is represented by a combination of values. These values define how the coefficients of the model have to be interpreted. By default, R uses treatment contrasts, as you can see when you check the relevant option like this:

```
> options("contrasts")
$contrasts
        unordered           ordered
 "contr.treatment"      "contr.poly"
```

Here you see that R uses different contrasts for unordered and ordered factors. These contrasts are actually contrast functions. They return a matrix with the contrast values for each level of the factor. The default contrasts for a factor with three levels look like this:

```
> X <- factor(c("A", "B", "C"))
> contr.treatment(X)
  B C
A 0 0
B 1 0
C 0 1
```

The two variables B and C are called that way because the variable B has a value of 1 if the factor level is B; otherwise, it has a value of 0. The same goes for C. Level A is represented by two zeros and called the *reference level.* In a one-factor model, the intercept is the mean of A.

You can change these contrasts using the same `options()` function, like this:

```
> options(contrasts = c("contr.sum", "contr.poly"))
```

The contrast function, `contr.sum()`, gives orthogonal contrasts where you compare every level to the overall mean. You can get more information about these contrasts on the Help page `?contr.sum`.

Evaluating the differences

To check the model, use the `summary()` function on the model object:

```
> summary(AOVModel)
          Df Sum Sq Mean Sq F value Pr(>F)
spray      5   2669   533.8    34.7 <2e-16 ***
Residuals 66   1015    15.4
---
Signif. codes:  0 '***' 0.001 '**' 0.01 '*' 0.05 '.' 0.1 ' ' 1
```

R prints the *analysis of variance table* that, in essence, tells you whether the different terms can explain a significant portion of the variance in your data. This table tells you only something about the term, but nothing about the differences between the different sprays. For that, you need to dig a bit deeper into the model.

Checking the model tables

With the `model.tables()` function, you look at the results for the individual levels of the factors. The function allows you to create two different tables; either with the estimated mean result for each group, or with the differences with the overall mean.

To know how much effect every spray had, you use the following code to see the differences with the overall mean:

```
> model.tables(AOVModel, type = "effects")
Tables of effects

 spray
spray
    A       B       C       D       E       F
 5.000   5.833  -7.417  -4.583  -6.000   7.167
```

Here you see that, for example, spray E resulted, on average, in six bugs fewer than the average over all fields. On the other hand, on fields where spray A was used, the farmers found, on average, five bugs more compared to the overall mean.

To get the modeled means per group and the overall mean, just use the argument value `type="means"` instead of `type="effects"`.

Looking at the individual differences

A farmer probably wouldn't consider buying spray A, but what about spray D? Although sprays E and C seem to be better, they also might be a lot more expensive. To test whether the pairwise differences between the sprays are significant, you use Tukey's Honest Significant Difference (HSD) test. The `TukeyHSD()` function allows you to do that very easily, like this:

```
> Comparisons <- TukeyHSD(AOVModel)
```

The `Comparisons` object now contains a list where every component is named after one factor in the model. In the example, you have only one component, called `spray`. This component contains, for every combination of sprays, the following:

✔ The difference between the means.

✔ The lower and upper level of the 95 percent confidence interval around that mean difference.

✔ The p-value that tells you whether this difference is significantly different from zero. This p-value is adjusted using the method of Tukey (hence, the column name `p adj`).

You can extract all that information using the classical methods for extraction. For example, you get the information about the difference between D and C like this:

```
> Comparisons$spray["D-C", ]
     diff        lwr        upr      p adj
2.8333333 -1.8660752  7.5327418  0.4920707
```

That difference doesn't look impressive, if you ask Tukey.

Plotting the differences

The `TukeyHSD` object has another nice feature: It can be plotted. Don't bother looking for a Help page of the plot function — all you find is one sentence: "There is a plot method." But it definitely works! Try it out like this:

```
> plot(Comparisons, las = 1)
```

You see the output of this simple line in Figure 15-4. Each line represents the mean difference between both groups with the according confidence interval. Whenever the confidence interval doesn't include zero (the vertical line), the difference between both groups is significant.

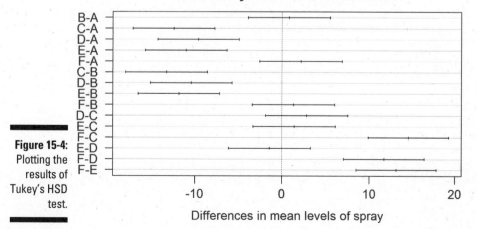

Figure 15-4: Plotting the results of Tukey's HSD test.

You can use some of the graphical parameters to make the plot more readable. Specifically, the las parameter is useful here. By setting las=1, you print all axis labels horizontally so you can actually read them. You can find out more about graphical parameters in Chapter 16.

Modeling linear relations

An analysis of variance also can be written as a *linear model,* where you use a factor as a predictor variable to model a response variable. In the previous section, you predict the mean bug count by looking at the insecticide that was applied.

Of course, predictor variables also can be continuous variables. For example, the weight of a car obviously has an influence on the mileage. But it would be nice to have an idea about the magnitude of that influence. Essentially, you want to find the equation that represents the trend line in Figure 15-5. You find the data you need for checking this in the dataset mtcars.

Building a linear model

The lm() function allows you to specify anything from the most simple linear model to complex interaction models. In this section, you build only a simple model to learn how to work with model objects.

To model the mileage in function of the weight of a car, you use the lm() function, like this:

```
> Model <- lm(mpg ~ wt, data = mtcars)
```

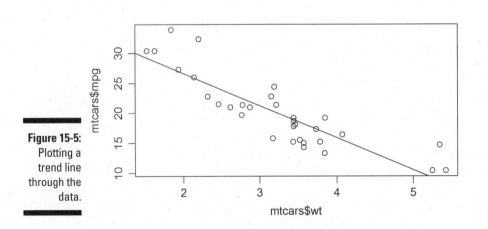

Figure 15-5:
Plotting a
trend line
through the
data.

You supply two arguments:

- ✔ **A formula that describes the model:** Here, you model the variable mpg as a function of the variable wt.

- ✔ **A data frame that contains the variables in the formula:** Here, you use the data frame mtcars.

You can specify many complex models with the formula interface when you know your way around. The "Details" section of the Help page ?formula provides all the information you need in great detail.

The resulting object is a list with a very complex structure, but in most cases you don't need to worry about that. The model object contains a lot of information that's needed for the calculations of diagnostics and new predictions.

Extracting information from the model

Instead of diving into the model object itself and finding the information somewhere in the list object, you can use some functions that help you to get the necessary information from the model. For example, you can extract a named vector with the coefficients from the model using the coef() function, like this:

```
> coef.Model <- coef(Model)
> coef.Model
(Intercept)           wt
  37.285126    -5.344472
```

These coefficients represent the intercept and the slope of the trend line in Figure 15-5. You can use this to plot the trend line on a scatterplot of the data. You do this in two steps:

1. **You plot the scatterplot with the data.**

   You use the plot() function for that. You discover more about this function in Chapter 16.

2. **You use the abline() function to draw the trend line based on the coefficients.**

The following code gives you the plot in Figure 15-5:

```
> plot(mpg ~ wt, data = mtcars)
> abline(coef = coef.Model)
```

The abline() argument coef takes a vector with two values: the intercept and the slope. That is exactly what the coef() function returns. Alternatively, you can specify the intercept and slope separately with the arguments a and b. You plot a vertical line by setting the argument v to the intercept with the *x*-axis instead. Horizontal lines are plotted by setting the argument h to the intercept with the *y*-axis.

In fact, the abline() function can also take a model object as argument. It uses the function coef() internally to look for the coefficients of the model. So in this case, you can further simplify the code and construct the regression line as follows:

```
> abline(Model)
```

In Table 15-1, you find an overview of functions to extract information from the model object itself. These functions work with different model objects, including those built by aov() and lm().

Many package authors also provide the same functions for the models built by the functions in their package. So, you can always try to use these extraction functions in combination with other model functions as well.

Table 15-1	Extracting Information from Model Objects
Function	*What It Does*
coef()	Returns a vector with the coefficients from the model
confint()	Returns a matrix with the upper and lower limit of the confidence interval for each coefficient of the model
fitted()	Returns a vector with the fitted values for every observation
residuals()	Returns a vector with the residuals for every observation
vcov()	Returns the variance-covariance matrix for the coefficient

Evaluating linear models

Naturally, R provides a whole set of different tests and measures not only to look at the model assumptions but also to evaluate how well your model fits your data. Again, the overview presented here is far from complete, but it gives you an idea of what's possible and a starting point to look deeper into the issue.

Summarizing the model

The `summary()` function immediately returns you the F test for models constructed with `aov()`. For `lm()` models, this is slightly different. Take a look at the output:

```
> Model.summary <- summary(Model)
> Model.summary

Call:
lm(formula = mpg ~ wt, data = mtcars)

Residuals:
    Min      1Q  Median      3Q     Max
-4.5432 -2.3647 -0.1252  1.4096  6.8727

Coefficients:
            Estimate Std. Error t value Pr(>|t|)
(Intercept)  37.2851     1.8776  19.858  < 2e-16 ***
wt           -5.3445     0.5591  -9.559 1.29e-10 ***
---
Signif. codes:  0 '***' 0.001 '**' 0.01 '*' 0.05 '.' 0.1 ' ' 1

Residual standard error: 3.046 on 30 degrees of freedom
Multiple R-squared: 0.7528,                Adjusted R-squared: 0.7446
F-statistic: 91.38 on 1 and 30 DF,  p-value: 1.294e-10
```

That's a whole lot of useful information. Here you see the following:

✔ The distribution of the residuals, which gives you a first idea about how well the assumptions of a linear model hold

✔ The coefficients accompanied by a t-test, telling you by how much every coefficient differs significantly from zero

✔ The goodness-of-fit measures R^2 and the adjusted R^2

✔ The F-test that gives you an idea about whether your model explains a significant portion of the variance in your data

You can use the `coef()` function to extract a matrix with the estimates, standard errors, t-value, and p-value for the coefficients from the summary object like this:

```
> coef(Model.summary)
             Estimate Std. Error   t value     Pr(>|t|)
(Intercept) 37.285126   1.877627 19.857575 8.241799e-19
wt          -5.344472   0.559101 -9.559044 1.293959e-10
```

If these terms don't tell you anything, look them up in a good source about modeling. For an extensive introduction to applying and interpreting linear models correctly, check out *Applied Linear Statistical Models,* 5th Edition, by Michael Kutner et al (McGraw-Hill/Irwin).

Testing the impact of model terms

To get an analysis of variance table — like the summary() function makes for an ANOVA model — you simply use the anova() function and pass it the lm() model object as an argument, like this:

```
> Model.anova <- anova(Model)
> Model.anova
Analysis of Variance Table

Response: mpg
          Df Sum Sq Mean Sq F value     Pr(>F)
wt         1 847.73  847.73  91.375 1.294e-10 ***
Residuals 30 278.32    9.28
---
Signif. codes:  0 '***' 0.001 '**' 0.01 '*' 0.05 '.' 0.1 ' ' 1
```

Here, the resulting object is a data frame that allows you to extract any value from that table using the subsetting and indexing tools from Chapter 7. For example, to get the p-value, you can do the following:

```
> Model.anova["wt", "Pr(>F)"]
[1] 1.293959e-10
```

You can interpret this value as the probability that adding the variable wt to the model doesn't make a difference. The low p-value here indicates that the weight of a car (wt) explains a significant portion of the difference in mileage (mpg) between cars. This shouldn't come as a surprise; a heavier car does, indeed, need more power to drag its own weight around.

The tests done by the anova() function use Type I (sequential) Sum of Squares, which is different from both SAS and SPSS. This also means that the order in which the terms are added to the model has an impact on the test values and the significance.

You can use the anova() function to compare different models as well, and many modeling packages provide that functionality. You find examples of this on most of the related Help pages like ?anova.lm and ?anova.glm.

Predicting new values

Apart from describing relations, models also can be used to predict values for new data. For that, many model systems in R use the same function, conveniently called `predict()`. Every modeling paradigm in R has a `predict()` function with its own flavor, but in general the basic functionality is the same for all of them.

Getting the values

For example, a car manufacturer has three designs for a new car and wants to know what the predicted mileage is based on the weight of each new design. In order to do this, you first create a data frame with the new values — for example, like this:

```
> new.cars <- data.frame(wt = c(1.7, 2.4, 3.6))
```

Always make sure the variable names you use are the same as used in the model. When you do that, you simply call the `predict()` function with the suited arguments, like this:

```
> predict(Model, newdata = new.cars)
        1        2        3
28.19952 24.45839 18.04503
```

So, the lightest car has a predicted mileage of 28.2 miles per gallon and the heaviest car has a predicted mileage of 18 miles per gallon, according to this model. Of course, if you use an inadequate model, your predictions will also be inadequate!

Having confidence in your predictions

In order to have an idea about the accuracy of the predictions, you can ask for intervals around your prediction. To get a matrix with the prediction and a 95 percent confidence interval around the mean prediction, you set the argument `interval` to `"confidence"`, like this:

```
> predict(Model, newdata = new.cars, interval = "confidence")
       fit      lwr      upr
1 28.19952 26.14755 30.25150
2 24.45839 23.01617 25.90062
3 18.04503 16.86172 19.22834
```

Now you know that — according to your model — a car with a weight of 2.4 tons has, *on average,* a mileage between 23 and 25.9 miles per gallon. In the same way, you can ask for a 95 percent prediction interval by setting the argument interval to "prediction":

```
> predict(Model, newdata = new.cars, interval = "prediction")
       fit      lwr      upr
1 28.19952 21.64930 34.74975
2 24.45839 18.07287 30.84392
3 18.04503 11.71296 24.37710
```

This information tells you that 95 percent of the cars with a weight of 2.4 tons have a mileage somewhere between 18.1 and 30.8 miles per gallon — assuming your model is correct, of course.

If you'd rather construct your own confidence interval, you can get the standard errors on your predictions as well by setting the argument se. fit to TRUE. You don't get a vector or a matrix; instead, you get a list with a component fit that contains the predictions and a component se.fit that contains the standard errors.

Part V
Working with Graphics

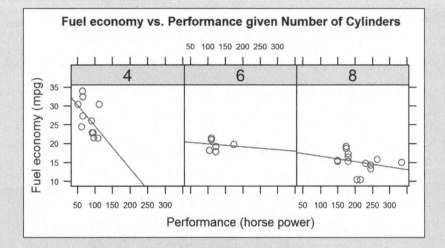

In this part . . .

✔ Introducing the essentials of graphical systems.

✔ Creating clear and elegant plots.

✔ Visit `www.dummies.com/extras/r` for great Dummies content online.

Chapter 16

Using Base Graphics

··

In This Chapter

▶ Creating a basic plot in R

▶ Changing the appearance of your plot

▶ Saving your plot as a picture

··

*I*n statistics and other sciences, being able to plot your results in the form of a graphic is often useful. An effective and accurate visualization can make your data come to life and convey your message in a powerful way.

R has very powerful graphics capabilities that can help you visualize your data. In this chapter, we give you a look at base graphics. It's called *base* graphics, because it's built into the standard distribution of R.

Creating Different Types of Plots

The base graphics function to create a plot in R is simply called plot(). This powerful function has many options and arguments to control all kinds of things, such as the plot type, line colors, labels, and titles.

The plot() function is a generic function (see Chapter 8), and R dispatches the call to the appropriate method. For example, if you make a scatterplot, R dispatches the call to plot.default(). The plot.default() function itself is reasonably simple and affects only the major look of the plot region and the type of plotting. All the other arguments that you pass to plot(), like colors, are used in internal functions that plot.default() simply happens to call.

Getting an overview of plot

To get started with plot, you need a set of data to work with. One of the built-in datasets is islands, which contains data about the surface area of the continents and some large islands on Earth.

First, create a subset of the ten largest islands in this dataset. You have many ways of doing this, but this line of code sorts islands in decreasing order, then uses head() to retrieve only the first ten elements:

```
> large.islands <- head(sort(islands, decreasing = TRUE), 10)
```

It is easy to create a plot with informative labels and titles. Try the following:

```
> plot(large.islands, main = "Land area of continents and islands",
+     ylab = "Land area in square miles")
> text(large.islands, labels = names(large.islands), adj = c(0.5, 1))
```

You can see the results in Figure 16-1. How does this work? The first line creates the basic plot with plot() and adds a main title and *y*-axis label. The second line adds text labels with the text() function. In the next section, you get to know each of these functions in more detail.

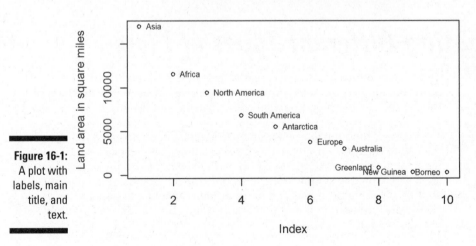

Figure 16-1: A plot with labels, main title, and text.

Adding points and lines to a plot

To illustrate some different plot options and types, look at the built-in dataset `faithful`. This is a data frame with observations of the eruptions of the Old Faithful geyser in Yellowstone National Park in the United States.

The built-in R datasets are documented in the same way as functions. So, you can get extra information on them by typing, for example, `?faithful`.

You've already seen that `plot()` creates a basic graphic. Try it with `faithful`:

```
> plot(faithful)
```

Figure 16-2 shows the resulting plot. Because `faithful` is a data frame with two columns, the plot is a scatterplot with the first column (eruptions) on the *x*-axis and the second column (waiting) on the *y*-axis.

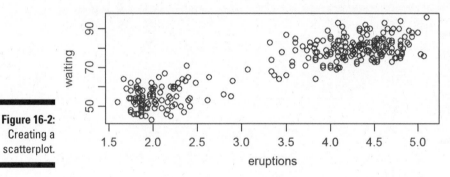

Figure 16-2: Creating a scatterplot.

Eruptions indicate the time in minutes for each eruption of the geyser, while waiting indicates the elapsed time between eruptions (also measured in minutes). As you can see from the general upward slope of the points, there tends to be a longer waiting period following longer eruptions.

Adding points

You add points to a plot with the `points()` function. You may have noticed that on the plot of `faithful` two clusters seem to be in the data. One cluster has shorter eruptions and waiting times — tending to last less than three minutes.

Create a subset of `faithful` containing eruptions shorter than three minutes:

```
> short.eruptions <- with(faithful, faithful[eruptions < 3, ])
```

Now use the `points()` function to add these points in red to your plot:

```
> plot(faithful)
> points(short.eruptions, col = "red", pch = 19)
```

You use the argument `col` to change the color of the points and the argument pch to change the plotting character. The value `pch=19` indicates a solid circle. To see all the arguments of `points()`, refer to `?points`.

Your resulting graphic should look like Figure 16-3, with the shorter eruption times indicated as solid red circles.

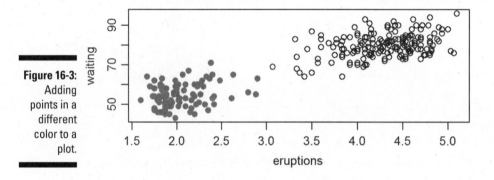

Figure 16-3:
Adding points in a different color to a plot.

Changing the shape of points

You've already seen that you can use the argument pch to change the plotting character when using points. This is described in more detail in the Help page for points, `?points`. For example, the Help page lists a variety of symbols, such as:

- ✔ **pch=19:** Solid circle
- ✔ **pch=20:** Bullet (smaller solid circle, two-thirds the size of `19`)
- ✔ **pch=21:** Filled circle
- ✔ **pch=22:** Filled square
- ✔ **pch=23:** Filled diamond
- ✔ **pch=24:** Filled triangle, point up
- ✔ **pch=25:** Filled triangle, point down

Changing the color

You can change the foreground (and where it makes sense also the background color) of symbols as well as lines. You've already seen how to set the foreground color using the argument `col="red"`. Some plotting symbols also use a background color, and you can use the argument `bg` to set the background color (for example, `bg="green"`). In fact, R has a number of predefined colors that you can use in graphics.

To get a list of available names for colors, you use the `colors()` function (or, if you prefer, `colours()`). The result is a vector of 657 elements with valid color names. Here are the first ten elements of this list:

```
> head(colors(), 10)
 [1] "white"         "aliceblue"     "antiquewhite"  "antiquewhite1"
 [5] "antiquewhite2" "antiquewhite3" "antiquewhite4" "aquamarine"
 [9] "aquamarine1"   "aquamarine2"
```

Adding lines to a plot

You add lines to a plot in a very similar way to adding points, except that you use the `lines()` function to achieve this.

But first, use a bit of R magic to create a trend line through the data, called a *regression model* (see Chapter 15). You use the `lm()` function to estimate a linear regression model:

```
fit <- lm(waiting ~ eruptions, data = faithful)
```

The result is an object of class `lm`. You use the function `fitted()` to extract the fitted values from a regression model (see Chapter 15). This is useful, because you can then plot the fitted values on a plot. You do this next.

To add this regression line to the existing plot, you simply use the function `lines()`. You also can specify the line color with the `col` argument:

```
> plot(faithful)
> lines(faithful$eruptions, fitted(fit), col = "blue")
```

The function `lines()` connects the data points in the exact order you give them. If you want to avoid your plot looking like a plate of spaghetti, you have to order your data points along the x axis. In the preceding example, this is already the case.

Another useful function is `abline()`. This allows you to draw horizontal, vertical, or sloped lines. To draw a vertical line at position `eruptions==3` in the color purple, try:

```
> abline(v = 3, col = "purple")
```

Your resulting graphic should look like Figure 16-4, with a vertical purple line at `eruptions==3` and a blue regression line.

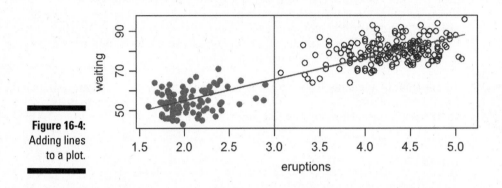

Figure 16-4:
Adding lines
to a plot.

To create a horizontal line, you also use `abline()`, but this time you specify the `h` argument. For example, create a horizontal line at the mean waiting time:

```
> abline(h = mean(faithful$waiting))
```

You also can use the function `abline()` to create a sloped line through your plot. In fact, by specifying the arguments `a` and `b`, you can draw a line that fits the mathematical equation `y = a + b*x`. In other words, if you specify the coefficients of your regression model as the arguments `a` and `b`, you get a line through the data that is identical to your prediction line:

```
> abline(a = coef(fit)[1], b = coef(fit)[2])
```

Even better, you can simply pass the `lm` object to `abline()` to draw the line directly. (This works because there is a method `abline.lm()`.) This makes your code very easy:

```
> abline(fit, col = "red")
```

Different plot types

The plot function has a `type` argument that controls the type of plot that gets drawn. For example, to create a plot with lines between data points, use `type="l"`; to plot only the points, use `type="p"`; and to draw both lines and points, use `type="b"`:

```
> plot(LakeHuron, type = "l", main = "type=\"l\"")
> plot(LakeHuron, type = "p", main = "type=\"p\"")
> plot(LakeHuron, type = "b", main = "type=\"b\"")
```

Note the use of the escape sequence \ " inside the string to indicate that you want a double quote in the main text. You encounter escaped quotes in Chapter 12.

Your resulting graphics should look similar to the three plots in Figure 16-5. The plot with lines only is on the left, the plot with points is in the middle, and the plot with both lines and points is on the right.

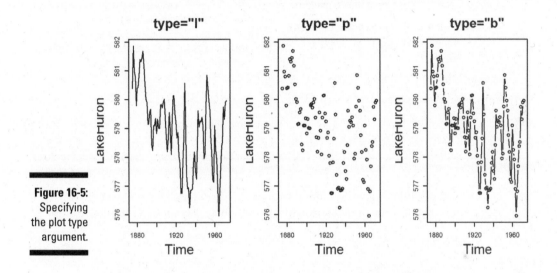

Figure 16-5: Specifying the plot type argument.

The Help page for plot() has a list of all the different types that you can use with the type argument:

- ✔ "p": Points
- ✔ "l": Lines
- ✔ "b": Both
- ✔ "c": The lines part alone of "b"
- ✔ "o": Both "overplotted"
- ✔ "h": Histogram like (or high-density) vertical lines
- ✔ "n": No plotting

It seems odd to use a plot function and then tell R not to plot it. But this can be very useful when you need to create just the titles and axes, and plot the data later using points(), lines(), or any of the other graphical functions.

Using R functions to create more types of plot

Aside from `plot()`, which gives you tremendous flexibility in creating your own plots, R also provides a variety of functions to make specific types of plots. (You use some of these in Chapters 14 and 15). Here are a few to explore:

✓ **Scatterplot:** If you pass two numeric vectors as arguments to `plot()`, the result is a scatterplot. Try:

```
> with(mtcars, plot(mpg, disp))
```

✓ **Box-and-whisker plot:** Use the `boxplot()` function:

```
> with(mtcars, boxplot(disp, mpg))
```

✓ **Histogram:** A histogram plots the frequency of observations. Use the `hist()` function:

```
> with(mtcars, hist(mpg))
```

✓ **Matrix of scatterplots:** The `pairs()` function is useful in data exploration, because it plots a matrix of scatterplots. Each variable gets plotted against another, as you see in Chapter 14:

```
> pairs(iris)
```

This flexibility may be useful if you want to build a plot step by step (for example, for presentations or documents). Here's an example:

```
> x <- seq(0.5, 1.5, 0.25)
> y <- rep(1, length(x))
> plot(x, y, type = "n")
> points(x, y)
```

In the next section, you take full control over the plot options and arguments, such as adding titles and labels or changing the font type of your plot.

Controlling Plot Options and Arguments

To really convey the message of your graphic, you may want to add titles and labels. You also can modify other elements of the graphic (for example, the type of box around the plot area or the font size of axis labels).

Base graphics allows you to take precise control over many plot options.

Adding titles and axis labels

You add the main title and axis labels with arguments to the `plot()` function:

- ✔ **main:** Main plot title
- ✔ **xlab:** *x*-axis label
- ✔ **ylab:** *y*-axis label

To add a title and axis labels to your plot of `faithful`, try the following:

```
> plot(faithful,
+     main = "Eruptions of Old Faithful",
+     xlab = "Eruption time (min)",
+     ylab = "Waiting time to next eruption (min)")
```

Your graphic should look like Figure 16-6.

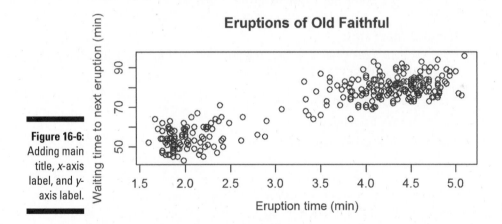

Figure 16-6:
Adding main
title, *x*-axis
label, and *y*-
axis label.

Changing plot options

You can change the look and feel of plots with a large number of options.

You can find all the documentation for changing the look and feel of base graphics in the Help page `?par`. This function allows you to set (or query) the graphical parameters or options.

Working with axes and legends

R allows you to also take control of other elements of a plot, such as axes, legends, and text:

✔ **Axes:** If you need to take full control of plot axes, use `axis()`. This function allows you to specify tickmark positions, labels, fonts, line types, and a variety of other options.

✔ **Legends:** You can use the `legend()` function to add legends, or keys, to plots.

✔ **Text:** In addition to legends, you can use the `text()` function to add text elements at any position on the plot.

The Help pages of the respective functions give you more information, and the examples contained in the Help pages show you how much you can do with these functions.

Notice that `par()` takes an extensive list of arguments. In this section, we describe a few of the most commonly used options.

The axes label style

To change the axes label style, use the graphics option `las` (label style). This changes the orientation angle of the labels:

✔ **0:** The default, parallel to the axis

✔ **1:** Always horizontal

✔ **2:** Perpendicular to the axis

✔ **3:** Always vertical

For example, to change the axis style to have all the axes text horizontal, use `las=1` as an argument to `plot`:

```
> plot(faithful, las = 1)
```

You can see what this looks like in Figure 16-7.

The box type

To change the type of box round the plot area, use the option `bty` (box type):

✔ **"o":** The default value draws a complete rectangle around the plot.

✔ **"n":** Draws nothing around the plot.

✔ **"l", "7", "c", "u", or "]":** Draws a shape around the plot area that resembles the uppercase letter of the option. So, the option `bty="l"` draws a line to the left and bottom of the plot.

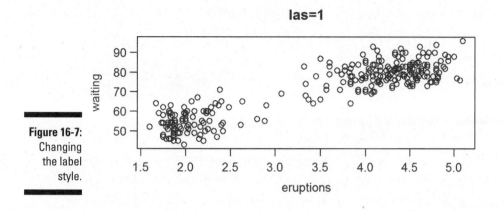

Figure 16-7:
Changing
the label
style.

To make a plot with no box around the plot area, use `bty="n"` as an argument to `plot`:

```
> plot(faithful, bty = "n")
```

Your graphic should look like Figure 16-8.

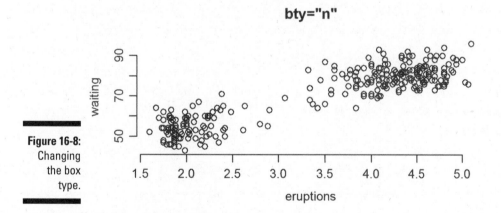

Figure 16-8:
Changing
the box
type.

More than one option

To change more than one graphics option in a single plot, simply add an additional argument for each plot option you want to set. For example, to change the label style, the box type, the color, and the plot character, try the following:

```
> plot(faithful, las = 1, bty = "l", col = "red", pch = 19)
```

The resulting plot is the plot in Figure 16-9.

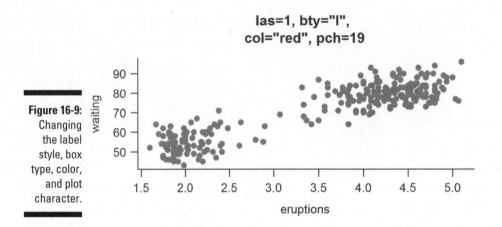

Font size of text and axes

To change the font size of text elements, use cex (short for character expansion ratio). The default value is 1. To reduce the text size, use a cex value of less than 1; to increase the text size, use a cex value greater than 1.

```
> x <- seq(0.5, 1.5, 0.25)
> y <- rep(1, length(x))
> plot(x, y, main = "Effect of cex on text size")
> text(x, y + 0.1, labels = x, cex = x)
```

Your plot should look like Figure 16-10 (left).

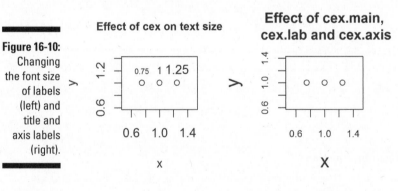

To change the size of other plot parameters, use the following:

- ✓ **cex.main:** Size of main title
- ✓ **cex.lab:** Size of axis labels (the text describing the axis)
- ✓ **cex.axis:** Size of axis text (the values that indicate the axis tick labels)

```
> plot(x, y, main = "Effect of cex.main, cex.lab and cex.axis",
+      cex.main = 1.25, cex.lab = 1.5, cex.axis = 0.75)
```

Your results should look like Figure 16-10 (right). Carefully compare the font size of the main title and the axes labels with the left side of Figure 16-10, and note how the main title as well as axes label fonts are larger while the axes annotations (tick labels) are smaller.

Putting multiple plots on a single page

To put multiple plots on the same graphics pages, you can use the graphics parameter mfrow or mfcol. To use this parameter, you need to supply a vector argument with two elements: the number of rows and the number of columns.

For example, to create two side-by-side plots, use mfrow=c(1, 2):

```
> old.par <- par(mfrow = c(1, 2))
> plot(faithful, main = "Faithful eruptions")
> plot(large.islands, main = "Islands", ylab = "Area")
> par(old.par)
```

When your plot is complete, you need to reset your par options. Otherwise, all your subsequent plots will appear side by side (until you close the active graphics device, or window, and start plotting in a new graphics device). We use a neat little trick to do this: When you make a call to par(), R sets your new options, but the return value from par() contains your old options. In the previous example, we save the old options to an object called old.par, and then reset the options after plotting using par(old.par).

Your result should look like Figure 16-11.

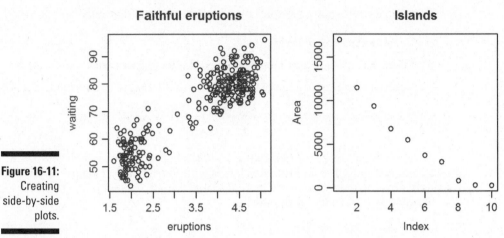

Figure 16-11:
Creating
side-by-side
plots.

Use `mfrow` to fill the plot grid by rows, and `mfcol` to fill the plot grid by columns. The Help page `?par` explains these options in detail, and also points you alternative layout mechanisms (like `layout()` or `split.screen()`).

If you want to put the graphics device back into a single plot state without using the `par(old.par)` trick, then use either

```
> par(mfcol = c(1, 1))
```

or

```
> par(mfrow = c(1, 1))
```

Because both these alternatives have the same effect, it does not matter which you use. If you work with the standard RGui, you can reset the graphics device by simply closing it. In both in R and RStudio, you also can use the function `dev.off()` to close (and hence reset) the graphics device.

Saving Graphics to Image Files

Much of the time, you may simply use R graphics in an interactive way to explore your data. But if you want to publish your results, you have to save your plot to a file and then import this graphics file into another document.

To save a plot to an image file, you have to do three things in sequence:

1. **Open a graphics device.**

   The default graphics device in R is your computer screen. To save a plot to an image file, you need to tell R to open a new type of device — in this case, a graphics file of a specific type, such as PNG, PDF, or JPG.

   The R function to create a PNG device is png(). Similarly, you create a PDF device with pdf() and a JPG device with jpg().

   If you are putting your graphics into a word processor, then PDF is often a good choice, because it creates highly detailed vector drawings. If you are putting your graphics onto a webpage or into a presentation, then PNG can be a good choice, because this creates small, compressed images.

2. **Create the plot.**

3. **Close the graphics device.**

   You do this with the dev.off() function.

Put this in action by saving a plot of faithful to the home folder on your computer. First set your working directory to your home folder (or to any other folder you prefer). If you use Linux or Mac, you'll be familiar with using "~/" as the shortcut to your home folder, but this also works on Windows:

```
> setwd("~/")
> getwd()
[1] "C:/Users/Andrie"
```

Next, write the three lines of code to save a plot to file:

```
> png(filename = "faithful.png")
> plot(faithful)
> dev.off()
```

Now you can check your file system to see whether the file faithful.png exists. (It should!) The result is a graphics file of type PNG that you can insert into a presentation, document, or website.

Chapter 17

Creating Faceted Graphics with Lattice

*C*reating subsets of data and plotting each subset allows you to see whether there are patterns between different subsets of the data. For example, a sales manager may want to see a sales report for different regions in the form of a graphic. A biologist may want to investigate different species of butterflies and compare the differences on a plot.

A single graphic that provides this kind of simultaneous view of different slices through the data is called a *faceted graphic*. Figure 17-1 shows a faceted plot of fuel economy and performance of motor cars. The important thing to notice is that the plot contains three panels, one each for cars with four, six, and eight cylinders.

R has a special package that allows you to easily create this kind of graphic. The package is called `lattice`, and in this chapter you get to draw `lattice` charts. Later in this chapter, you create a `lattice` plot that should be identical to Figure 17-1.

In this chapter, we give the briefest of introductions to the extensive functionality in `lattice`. An entire book could be written about `lattice` graphics — and, in fact, such a book already exists. The author of the `lattice` package, Deepayan Sarkar, also wrote a book called *Lattice: Multivariate Data Visualization with R* (Springer). You can find the figures and code from that book at `http://lmdvr.r-forge.r-project.org/figures/figures.html`.

Fuel economy vs. Performance given Number of Cylinders

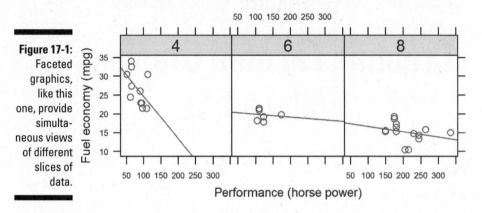

Figure 17-1:
Faceted
graphics,
like this
one, provide
simulta-
neous views
of different
slices of
data.

Creating a Lattice Plot

To explore lattice graphics, first take a look at the built-in dataset mtcars. This dataset contains 32 observations of motor cars and information about the engine, such as number of cylinders, automatic versus manual gearbox, and engine power.

All the built-in datasets of R also have good help information that you can access through the Help mechanism — for example, by typing **?mtcars** into the R console:

```
> str(mtcars)
'data.frame': 32 obs. of  11 variables:
 $ mpg : num  21 21 22.8 21.4 18.7 18.1 14.3 24.4 22.8 19.2 . . .
 $ cyl : num  6 6 4 6 8 6 8 4 4 6 . . .
 $ disp: num  160 160 108 258 360 . . .
 $ hp  : num  110 110 93 110 175 105 245 62 95 123 . . .
 ....
 $ carb: num  4 4 1 1 2 1 4 2 2 4 ..
```

Say you want to explore the relationship between fuel economy, engine power, and number of cylinders. The mtcars dataset has three elements with this information:

✔ **mpg:** Fuel economy measured in miles per gallon (mpg)

✔ **hp:** Engine power measured in horsepower (hp)

✔ **cyl:** Number of cylinders

In this section, you create different plots of mpg against hp and cyl.

However, you first have to do some data cleanup. The variable `cyl` is numeric, when really it is a categorical variable and should be a factor. Further in this chapter, you also need the variable `am`, a factor variable with value 0 for automatic and 1 for manual gearbox. Finally, you want to store the names of the cars in a proper variable instead of the rownames.

To prepare the data for plotting, first create the new transformed object:

```
> transform.mtcars <- transform(mtcars,
+       cyl = factor(cyl),
+       am = factor(am, labels = c("Automatic", "Manual")),
+       cars = rownames(mtcars)
+ )
```

Loading the lattice package

Although the `lattice` package forms part of the R distribution, you have to tell R that you plan to use the code in this package. You do this with the `library()` function. Remember that you need to do this at the start of each clean R session in which you want to use `lattice`:

```
> library("lattice")
```

Making a lattice scatterplot

The `lattice` package has a number of different functions to create different types of plot. For example, to create a scatterplot, use the `xyplot()` function. Notice that this is different from base graphics, where the `plot()` function creates a variety of different plot types (because of the method dispatch mechanism). Besides `xyplot()`, we briefly discuss the other `lattice` functions later in this chapter.

To make a `lattice` plot, you need to specify at least two arguments:

- ✔ **formula:** This is a formula typically of the form $y \sim x|z$. It means to create a plot of y against x, conditional on z. In other words, create a plot for every unique value of z. Each of the variables in the `formula` has to be a column in the data frame that you specify in the `data` argument.

- ✔ **data:** A data frame that contains all the columns that you specify in the `formula` argument.

This example should make it clear:

```
> xyplot(mpg ~ hp | cyl, data = transform.mtcars)
```

You can see that

✔ The variables mpg, hp, and cyl are columns in the data frame transform.mtcars.

✔ Although cyl is a numeric vector in the original mtcars, the number of cylinders in a car can be only whole numbers (or *discrete variables,* in statistical jargon). By using factor(cyl) to create transform.mtcars, you tell R that cyl is, in fact, a discrete variable. If you forget to do this, R will still create a graphic, but the labels of the strips at the top of each panel will be displayed differently.

Your code should produce a graphic that looks like Figure 17-2. Because each of the cars in the data frame has four, six, or eight cylinders, the chart has three panes. You can see that the cars with larger engines tend to have more power (hp) and poorer fuel consumption (mpg).

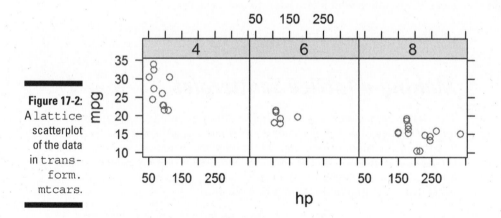

Figure 17-2: A lattice scatterplot of the data in transform.mtcars.

Adding trend lines

In Chapter 15, we show you how to create *trend lines,* or regression lines through data.

When you tell lattice to calculate a line of best fit, it does so for each panel in the plot. This is straightforward using xyplot(), because it's as simple as adding a type argument. In particular, you want to specify that the type is both points (type = "p") and regression (type = "r"). You can combine different types with the c() function, like this:

```
> xyplot(mpg ~ hp | cyl, data = transform.mtcars,
+        type = c("p", "r"))
```

Your graphic should look like Figure 17-3.

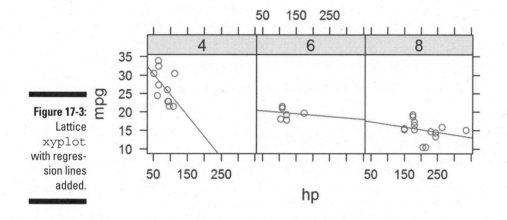

Figure 17-3: Lattice `xyplot` with regression lines added.

 Strictly speaking, type is not an argument to xyplot(), but an argument to panel.xyplot(). You can control the panels of lattice graphics with a panel function. The function xyplot() calls this panel function internally, using the type argument you specified. The default panel function for xyplot() is panel.xyplot(). Similarly, the panel function for barchart() — which we cover later in this chapter — is panel.barchart(). The panel function allows you to take fine control over many aspects of your chart. You can find out more in the excellent Help for these functions — for example, by typing **?panel.xyplot** into your R console.

Base, grid, and lattice graphics

Perhaps confusingly, the standard distribution of R actually contains three different graphics packages:

✔ **Base graphics** is the graphics system that was originally developed for R. The workhorse function of base graphics is plot()

(see Chapter 16). The code for base graphics is in the graphics package, which is loaded by default when you start R.

✔ **Grid graphics** is an alternative graphics system that was later added to R. The big difference between grid and the original

(continued)

(continued)

base graphics system is that grid allows for the creation of multiple regions, called *viewports,* on a single graphics page. Grid is a framework of code and doesn't, by itself, create complete charts. The author of grid, Paul Murrell, describes some of the ideas behind grid graphics on his website at `www.stat.auckland.ac.nz/~paul/grid/doc/grid.pdf`. *Note:* The grid package needs to be loaded before you can use it.

✔ **Lattice** is a graphics system that specifically implements the idea of Trellis graphics (or faceted graphics), which was originally developed for the languages S and S-Plus at Bell Labs. Lattice graphics in R make use of grid graphics. This means that the functions for creating graphics and changing options in `base` and `lattice` are mostly incompatible with one another. The `lattice` package needs to be loaded before use.

Changing Plot Options

R has a very good reputation for being able to create publication-quality graphics. If you want to use your `lattice` graphics in reports or documents, you'll probably want to change the plot options.

The `lattice` package makes use of the grid graphics engine, which is completely different from the base graphics in Chapter 16. Because of this, none of the mechanisms for changing plot options covered in Chapter 16 are applicable to `lattice` graphics.

Adding titles and labels

To add a main title and axis labels to a `lattice` plot, you can specify the following arguments:

✔ **main:** Main title

✔ **xlab:** *x*-axis label

✔ **ylab:** *y*-axis label

```
> xyplot(mpg ~ hp | cyl, data = transform.mtcars,
+       type = c("p", "r"),
+       main = "Fuel economy vs. Performance",
+       xlab = "Performance (horse power)",
+       ylab = "Fuel economy (mpg)",
+ )
```

Your output should now be similar to Figure 17-4.

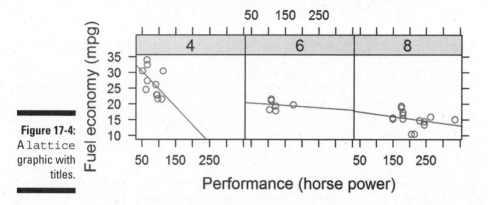

Fuel economy vs. Performance

Changing the font size of titles and labels

You probably think that the title and label text in Figure 17-4 are disproportionately large compared to the rest of the graphic.

To change the size of your labels, you need to modify your arguments to be lists. Similar to base graphics, you specify a cex argument in lattice graphics to modify the character expansion ratio. For example, to reduce the main title and axis label text to 75 percent of standard size, specify cex=0.75 as an element in the list argument to main, xlab, and ylab.

To keep it simple, build up the formatting of your plot step by step. Start by changing the size of your main title to cex=0.75:

```
> xyplot(mpg ~ hp | cyl, data = transform.mtcars,
+       type = c("p", "r"),
+       main = list(
+         label = "Fuel economy vs. Performance given Number of Cylinders",
+         cex = 0.75)
+ )
```

Do you see what happened? Your argument to main now contains a list with two elements: label and cex.

You construct the arguments for `xlab` and `ylab` in exactly the same way. Each argument is a list that contains the label and any other formatting options you want to set. Expand your code to modify the axis labels:

```
> xyplot(mpg ~ hp | cyl, data = transform.mtcars,
+        type = c("p", "r"),
+        main = list(
+            label = "Fuel economy vs. Performance given Number of Cylinders",
+            cex = 0.75),
+        xlab = list(
+            label = "Performance (horse power)",
+            cex = 0.75),
+        ylab = list(
+            label = "Fuel economy (mpg)",
+            cex = 0.75),
+        scales = list(cex = 0.5)
+ )
```

If you look carefully, you'll see that the code includes an argument to modify the size of the scales text to 50 percent of standard (`scales = list(cex = 0.5)`). Your results should look like Figure 17-5.

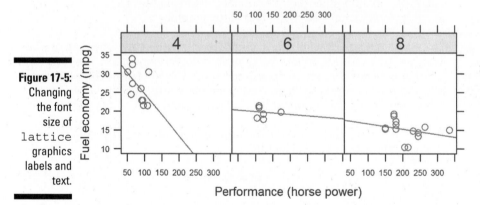

Figure 17-5: Changing the font size of `lattice` graphics labels and text.

Using themes to modify plot options

One neat feature of `lattice` graphics is that you can create themes to change the plot options of your charts. To do this, you need to use the `par.settings` argument. In Chapter 16, you use the `par()` function to update graphics parameters of base graphics. The `par.settings` argument in `lattice` is similar.

The easiest way to use the par.settings argument is to use it in conjunction with the simpleTheme() function. With simpleTheme(), you can specify the arguments for the following:

- ✔ **col, col.points, col.line:** Control the colors of symbols, points, lines, and other graphics elements such as polygons

- ✔ **cex, pch, font:** Control the character expansion ratio (cex), plot character (pch), and font type

- ✔ **lty, lwd:** Control the line type and line width

For example, to modify your plot to have red points and a blue regression line, use the following:

```
> xyplot(mpg ~ hp | cyl, data = transform.mtcars,
+        type = c("p", "r"),
+        par.settings = simpleTheme(col = "red", col.line = "blue")
+ )
```

You can see the result in Figure 17-6.

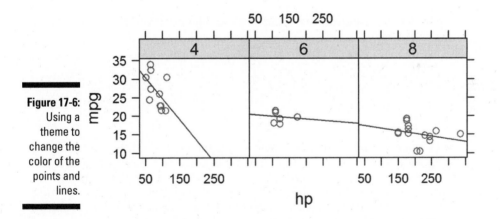

Figure 17-6: Using a theme to change the color of the points and lines.

Plotting Different Types

With lattice graphics, you can create many different types of plots, such as scatterplots and bar charts. Here are just a few of the different types of plots you can create:

- ✔ **Scatterplot:** xyplot()
- ✔ **Bar chart:** barchart()
- ✔ **Box-and-whisker plot:** bwplot()

- ✔ **One-dimensional strip plot:** `stripplot()`
- ✔ **Three-dimensional scatterplots:** `cloud()`
- ✔ **Three-dimensional surface plots:** `wireframe()`

For a complete list of the different types of `lattice` plots, see the Help at `?lattice`.

Because making bar charts and making box-and-whisker plots are such common activities, we discuss these functions in the following sections.

Making a bar chart

To make a bar chart, use the `lattice` function `barchart()`. Say you want to create a bar chart of fuel economy for each different type of car. To do this, you first have to add the names of the cars to the data itself. Because the names are contained in the row names, this means assigning a new column in your data frame with the name `cars`, containing `rownames(mtcars)`. You already did this earlier in the chapter, when you created `transform.mtcars`.

Now you can create your bar chart using similar syntax to the scatterplot you made earlier:

```
> barchart(cars ~ mpg | cyl, data = transform.mtcars,
+          main = "barchart",
+          scales = list(cex = 0.5),
+          layout = c(3, 1)
+ )
```

Once again (because you have eagle eyes), you've noticed the additional argument `layout` in this code. Lattice plots adapt to the size of the active graphics window on your screen. They do this by changing the configuration of the panels of your plot. For example, if your graphics window is too narrow to contain the panels side by side, then `lattice` will start to stack your panels.

You control the layout of your panels with the argument `layout`, consisting of two numbers indicating the number of columns and number of rows in your plot. In our example, we want to ensure that the three panels are side by side, so we specify `layout=c(3, 1)`.

Your plot should look like Figure 17-7.

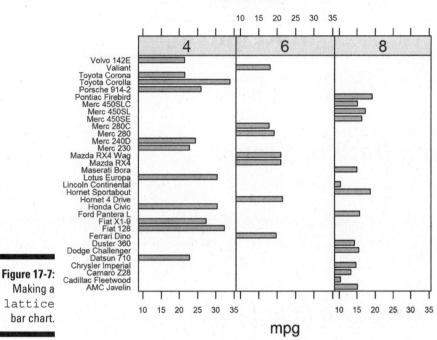

Figure 17-7:
Making a
`lattice`
bar chart.

Making a box-and-whisker plot

A box-and-whisker plot is useful when you want to visually summarize the uncertainty of a variable. The plot consists of a dark circle at the mean; a box around the upper and lower hinges (the hinges are at approximately the 25th and 75th percentiles); and a dotted line, or whisker, at most 1.5 times the box length (see Chapter 14).

The `lattice` function to create a box-and-whisker plot is `bwplot()`, and you can see the result in Figure 17-8.

Notice that the function formula does not have a left-hand side to the equation. Because you're creating a one-dimensional plot of horsepower conditional on cylinders, the formula simplifies to ~ hp | cyl. In other words, the formula starts with the tilde symbol:

```
> bwplot(~ hp | cyl, data = transform.mtcars, main = "bwplot")
```

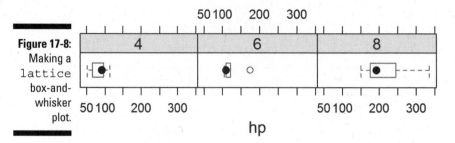

Figure 17-8:
Making a
`lattice`
box-and-
whisker
plot.

Plotting Data in Groups

Often, you want to create plots where you compare different groups in your data. In this section, you first take a look at data in tall format as opposed to data in wide format. When you have data in tall format, you can easily use `lattice` graphics to visualize subgroups in your data. Then you create some charts with contained subgroups. Finally, you add a key, or legend, to your plot to indicate the different subgroups.

Using data in tall format

So far, you've graphed only one variable against another in your `lattice` plots. In most of the examples, you plotted `mpg` against `hp` for each unique value of `cyl`. But what happens when you want to analyze more than one variable simultaneously?

Consider the built-in dataset `longley`, containing data about employment, unemployment, and other population indicators:

```
> str(longley)
'data.frame':  16 obs. of  7 variables:
 $ GNP.deflator: num  83 88.5 88.2 89.5 96.2 . . .
 $ GNP         : num  234 259 258 285 329 . . .
 $ Unemployed  : num  236 232 368 335 210 . . .
 $ Armed.Forces: num  159 146 162 165 310 . . .
 $ Population  : num  108 109 110 111 112 . . .
 $ Year        : int  1947 1948 1949 1950 1951 1952 1953 1954 1955 1956 . . .
 $ Employed    : num  60.3 61.1 60.2 61.2 63.2 . . .
```

One way to easily analyze the different variables of a data frame is to first reshape the data frame from wide format to tall format.

A wide data frame contains a column for each variable (see Chapter 13). A tall data frame contains all the same information, but the data is organized in such a way that one column is reserved for identifying the name of the variable and a second column contains the actual data.

An easy way to reshape a data frame from wide format to tall format is to use the `melt()` function in the `reshape2` package. ***Remember:*** `reshape2` is not part of base R — it's an add-on package that is available on CRAN. You can install it with the `install.packages("reshape2")` function.

```
> library("reshape2")
> mlongley <- melt(longley, id.vars = "Year")
> str(mlongley)
'data.frame': 96 obs. of  3 variables:
 $ Year    : int  1947 1948 1949 1950 1951 1952 1953 1954 1955 1956 . . .
 $ variable: Factor w/ 6 levels "GNP.deflator",..: 1 1 1 1 1 1 1 1 1 1 . . .
 $ value   : num  83 88.5 88.2 89.5 96.2 . . .
```

Now you can plot the tall data frame `mlongley` and use the new columns `value` and `variable` in the formula `value~Year | variable`:

```
> xyplot(value ~ Year | variable, data = mlongley,
+        layout = c(6, 1),
+        par.strip.text = list(cex = 0.5),
+        scales = list(cex = 0.5)
+ )
```

The additional arguments `par.strip.text` and `scales` control the font size (character expansion ratio) of the strip at the top of the chart, as well as the scale, as you can see in Figure 17-9.

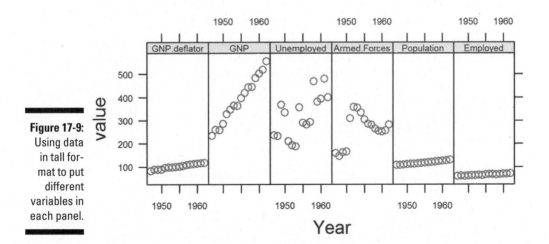

Figure 17-9: Using data in tall format to put different variables in each panel.

When you create plots with multiple groups, make sure that the resulting plot is meaningful. For example, Figure 17-9 plots the `longley` data, but it can be misleading because the units of measurement are very different. For example, the unit of GNP (short for *Gross National Product*) is probably billions of dollars. In contrast, the unit of population is probably millions of people. (The documentation of the `longley` dataset is not clear on this topic.) Be very careful when you present plots like this — you don't want to be accused of creating *chart junk* (misleading graphics).

Creating a chart with groups

Many graphics types — but bar charts in particular — tend to display multiple groups of data at the same time. Usually, you can distinguish different groups by their color or sometimes their shading.

If you ever want to add different colors to your plot to distinguish between different data, you need to define groups in your `lattice` plot.

Say you want to create a bar chart that differentiates whether a car has an automatic or manual gearbox. The `transform.mtcars` dataset has a column with this data, called `am`. (Remember you transformed this variable to a factor in the beginning of this chapter.) You plot your data using the same formula as before, but this time you add an argument defining the group, `group=am`:

```
> barchart(cars ~ mpg | cyl, data = transform.mtcars,
+          group = am,
+          scales = list(cex = 0.5),
+          layout = c(3, 1),
+ )
```

When you run this code, you'll get your desired bar chart. However, the first things you'll notice are that the colors look a bit washed out and you don't have a key to distinguish between automatic and manual cars.

Adding a key

It is easy to add a key to a graphic that already contains a `group` argument. Usually, it's as simple as adding another argument, `auto.key = TRUE`, which automatically creates a key that matches the groups:

```
> plot.colours <- c("grey80", "grey20")
> barchart(cars ~ mpg | cyl, data = transform.mtcars,
+    main = "barchart with groups",
```

```
+      group = am,
+      auto.key = TRUE,
+      par.settings = simpleTheme(col = plot.colours, border = plot.colours),
+          scales = list(cex = 0.5),
+          layout = c(3, 1)
+ )
```

One more thing to notice about this specific example is the arguments for
`par.settings` to control the color of the bars. In this case, the colors are
shades of gray. You can see the effect in Figure 17-10.

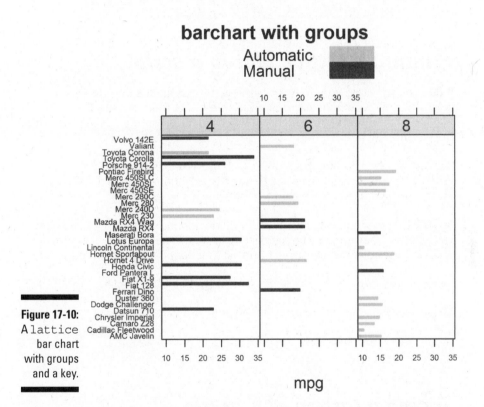

Figure 17-10:
A `lattice`
bar chart
with groups
and a key.

Printing and Saving a Lattice Plot

You need to know three other essential things about `lattice` plots: how to
assign a `lattice` plot to an object, how to print a `lattice` plot in a script,
and how to save a `lattice` plot to file. That's what we cover in this section.

Assigning a lattice plot to an object

Lattice plots are objects; therefore, you can assign them to variables, just like any other object. This is very convenient when you want to reuse a plot object in your downstream code — for example, to print it later.

The assignment to a variable works just like any variable assignment in R:

```
> my.plot <- xyplot(mpg ~ hp | cyl, data = transform.mtcars)
> class(my.plot)
[1] "trellis"
```

Printing a lattice plot in a script

When you run code interactively — by typing commands into the R console — simply typing the name of a variable prints that variable. However, you need to *explicitly* print an object when running a script. You do this with the print() function.

Because a lattice plot is an object, you need to explicitly use the print() function in your scripts. This is a frequently asked question in the R documentation, and it can easily lead to confusion if you forget.

To be clear, the following line of code will do *nothing* if you put it in a script and source the script. (To be technically correct: The code will still run, but the resulting object will never get printed — it simply gets discarded.)

```
> xyplot(mpg ~ hp | cyl, data = transform.mtcars)
```

To get the desired effect of printing the plot, you must use print():

```
> my.plot <- xyplot(mpg ~ hp | cyl, data = transform.mtcars)
> print(my.plot)
```

Saving a lattice plot to file

To save a lattice plot to an image file, you use a slightly modified version of the sequence of functions that you came across in base graphics (see Chapter 16).

Here's a reminder of the sequence:

1. **Open a graphics device using, for example, png().**

   Tip: The lattice package provides the trellis.device() function that effectively does the same thing, but it's optimized for lattice plots, because it uses appropriate graphical parameters.

2. **Print the plot.**

   Remember: You must use the print() function explicitly!

3. **Close the graphics device.**

Put this into action using trellis.device() to open a file called xyplot.png, print your plot, and then close the device. (You can use setwd("~/") to set your working directory to your home folder; see Chapter 16.)

```
> setwd("~/")
> trellis.device(device = "png", filename = "xyplot.png")
> print(my.plot)
> dev.off()
```

You should now be able to find the file xyplot.png in your home folder.

To change the working directory and then change it back to the original, you would do something like:

```
origwd <- getwd()
setwd("~/") # Set the working directory here
# do stuff
setwd(origwd)
```

Chapter 18

Looking At ggplot2 Graphics

In This Chapter

▶ Installing and loading the ggplot2 package

▶ Understanding how to use build a plot using layers

▶ Creating charts with suitable geoms and stats

▶ Adding facets to your plot

One of the strengths of R is that it's more than just a programming language — it also has thousands of packages written and contributed by independent developers. One of these packages, ggplot2, is tremendously popular and offers a new way of creating insightful graphics using R.

Much of the ggplot2 philosophy is based on the so-called "grammar of graphics," a consistent sound way of describing all the components that go into a graphical plot. You don't need to know anything about the grammar of graphics to use ggplot2 effectively, but now you know where its name comes from.

In this chapter, you first install and load the ggplot2 package and then take a first look at layers, the building blocks of the ggplot2 graphics. Next, you define the data, geoms, and stats that make up a layer, and use these to create some plots. Finally you take full control over your graphics by adding facets and scales as well as controlling other plot options, such as adding labels and titles.

Installing and Loading ggplot2

Because ggplot2 isn't part of the standard distribution of R, you have to download the package from CRAN and install it.

In Chapter 3, you see that the Comprehensive R Archive Network (CRAN) is a network of servers around the world that contain the source code, documentation and add-on packages for R. Its home page is at `http://cran.r-project.org`.

Each submitted package on CRAN also has a page that describes what the package is about. You can view the `ggplot2` page at `http://cran.r-project.org/web/packages/ggplot2/index.html`.

Although it's fairly common practice to simply refer to the package as `ggplot`, it is, in fact, the second implementation of the grammar of graphics for R; hence, the package is `ggplot2`. Version 1.0.0 of `ggplot2` was released in May 2014.

Perhaps somewhat confusingly, the most important function in this package is `ggplot()`. Notice that the function doesn't have a 2 in its name. So, be careful to include the 2 when you refer to the package in your R code (for example, when using `install.packages()` or `library()`), but remember that the *function* `ggplot()` itself does not contain a 2.

In Chapter 3, you also see how to install a package for the first time with the `install.packages()` function and to load the package at the start of each R session with the `library()` function.

To install the `ggplot2` package, use:

```
> install.packages("ggplot2")
```

And then to load the package, use:

```
> library("ggplot2")
```

Looking At Layers

The basic concept of a `ggplot2` graphic is that you combine different plot elements into *layers*. Each layer of a `ggplot2` graphic contains information about the following:

- **The data that you want to plot:** For `ggplot()`, this must be a data frame.

- **A mapping from the data to your plot:** This usually is as simple as telling `ggplot()` what goes on the *x*-axis and what goes on the *y*-axis. (In the "Mapping data to plot aesthetics" section, later in this chapter, we explain how to use the `aes()` function to set up the mapping.)

✔ **A geometric object, or geom in** `ggplot` **terminology:** The geom defines the overall look of the layer (for example, whether the plot is made up of bars, points, or lines).

✔ **A statistical summary, called a stat in** `ggplot`**:** This describes how you want the data to be summarized (for example, binning for histograms, or smoothing to draw regression lines). Note that each stat has a default associated geom, and vice versa.

That was a mouthful. In practice, you describe all this in a short line of code. For example, here is the ggplot2 code to plot the `faithful` data using two layers. (Because you plot `faithful` in Chapter 16, we won't bore you by describing it here.) The first layer is a geom that draws the points of a scatterplot; the second layer is a stat that draws a smooth line through the points.

```
> ggplot(faithful, aes(x = eruptions, y = waiting)) +
+    geom_point() +
+    stat_smooth()
```

This single line of code creates the graphic in Figure 18-1.

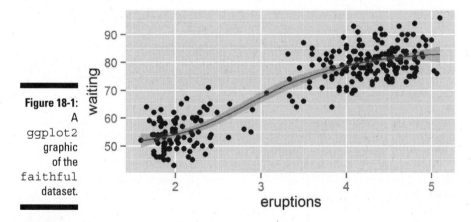

Figure 18-1:
A
ggplot2
graphic
of the
`faithful`
dataset.

The ggplot2 package is generous with tips in the form of messages, as you can see when running the preceding code. Unless these messages mention the words *warning* or *error,* you can safely ignore them.

Using Geoms and Stats

To create a ggplot2 graphic, you have to explicitly tell the function what's in each of the components of the layer. In other words, you have to tell the `ggplot()` function your data, the mapping between your data and the geom, and then either a geom or a stat. Since each geom has a default stat, and each stat has a default geom, you don't have to specify both, unless you want to take full control of the plot.

In this section, we discuss geoms and then stats.

Defining what data to use

The first element of a `ggplot2` layer is the data. There is only one rule for supplying data to `ggplot()`: Your data must be in the form of a data frame. This is different from base graphics, which allow plotting of data in vectors, matrices, and other structures.

In the remainder of this chapter, we use the built-in dataset `quakes`. This dataset is a data frame with information about earthquakes near Fiji.

You tell `ggplot()` what data to use and how to map your data to your geom in the `ggplot()` function. The `ggplot()` function takes two arguments:

- ✔ **data:** a data frame with your data (for example, `data=quakes`).
- ✔ **...:** The dots argument indicates that any other argument you specified here gets passed on to downstream functions (that is, other functions that `ggplot()` happens to call). In the case of `ggplot()`, this means that anything you specify in this argument is available to your geoms and stats that you define later.

Because the dots argument is available to any geom or stat in your plot, it's a convenient place to define the mapping between your data and the visual elements of your plot.

This is where you typically specify a mapping between your data and your geom.

Mapping data to plot aesthetics

After you've told `ggplot()` what data to use, the next step is to tell it how your data corresponds to visual elements of your plot. This mapping between data and visual aesthetics is the second element of a `ggplot2` layer.

The visual elements of a plot, or *aesthetics,* include lines, points, symbols, colors, position . . . anything that you can see. For example, you can map a column of your data to the *x*-axis of your plot, or you can map a column of your data to correspond to the *y*-axis of your plot. You also can map data to groups, colors, or the size of points in scatterplots — in fact, you can map your data to anything that your geom supports.

You use the special function aes() to set up a mapping between data and aesthetics. Each argument to aes() maps a column in your data to a specific element in your geom.

Take another look at the code used to create Figure 18-1:

```
> ggplot(faithful, aes(x = eruptions, y = waiting)) +
+    geom_point() +
+    stat_smooth()
```

You can see that this code tells ggplot() to use the data frame faithful as the data source. And now you understand that aes() creates a mapping between the *x*-axis and faithful$eruptions, as well as between the *y*-axis and faithful$waiting.

The next thing you notice about this code is the plus (+) signs at the end of each line. In ggplot2, you use the + operator to combine the different layers of the plot.

In summary, you use the aes() function to define the mapping between your data and your plot. This is simple enough, but it leaves one question: How do you know which aesthetics are available in different geoms?

Getting geoms

A ggplot2 geom tells the plot how you want to display your data. For example, you use geom_bar() to make a bar chart. In ggplot2, you can use a variety of predefined geoms to make standard types of plot.

A geom defines the layout of a ggplot2 layer. For example, you can use geoms to create bar charts, scatterplots, and line diagrams (as well as a variety of other plots), as you can see in Table 18-1.

Each geom has a default stat, and each stat has a default geom. In practice, you have to specify only one of these.

Table 18-1   A Selection of Geoms and Associated Default Stats

Geom	Description	Default Stat
`geom_bar()`	Bar chart	`stat_bin()`
`geom_point()`	Scatterplot	`stat_identity()`
`geom_line()`	Line diagram, connecting observations in order by *x*-value	`stat_identity()`
`geom_boxplot`	Box-and-whisker plot	`stat_boxplot()`
`geom_path`	Line diagram, connecting observations in original order	`stat_identity()`
`geom_smooth`	Add a smoothed conditioned mean	`stat_smooth()`
`geom_histogram`	An alias for `geom_bar()` and `stat_bin()`	`stat_bin()`

Creating a bar chart

To make a bar chart, you use the `geom_bar()` function. However, note that the default stat is `stat_bin()`, which is used to cut your data into bins. Thus, the default behavior of `geom_bar()` is to create a histogram.

For example, to create a histogram of the depth of earthquakes in the `quakes` dataset, try:

```
> ggplot(quakes, aes(x = depth)) + geom_bar()
> ggplot(quakes, aes(x = depth)) + geom_bar(binwidth = 50)
```

Notice that your mapping defines only the *x*-axis variable (in this case, `quakes$depth`). A useful argument to `geom_bar()` is `binwidth`, which controls the size of the bins that your data is cut into. This creates the plot of Figure 18-2.

So, if `geom_bar()` makes a histogram by default, how do you make a bar chart? The answer is that you first have to aggregate your data, and then specify the argument `stat="identity"` in your call to `geom_bar()`.

In the next example, you use `aggregate()` (see Chapter 13) to calculate the number of quakes at different depth strata:

```
> quakes.agg <- aggregate(mag ~ round(depth, -1), data = quakes,
+                         FUN = length)
> names(quakes.agg) <- c("depth", "mag")
```

Now you can plot the object `quakes.agg` with `geom_bar(stat="identity")`:

```
> ggplot(quakes.agg, aes(x = depth, y = mag)) +
+   geom_bar(stat = "identity")
```

Your results should be very similar to Figure 18-2.

In summary, you can use `geom_bar()` to create a histogram and let `ggplot2` summarize your data, or you can summarize your data and then use `stat="identity"` to plot a bar chart.

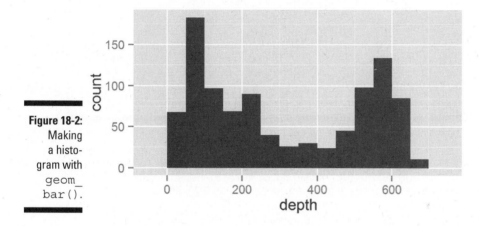

Figure 18-2: Making a histogram with `geom_bar()`.

Making a scatterplot

To create a scatterplot, you use the `geom_point()` function. A scatterplot creates points (or sometimes bubbles or other symbols) on your chart. Each point corresponds to an observation in your data.

You've probably seen or created this type of graphic a million times, so you already know that scatterplots use the Cartesian coordinate system, where one variable is mapped to the *x*-axis and a second variable is mapped to the *y*-axis.

In exactly the same way, in `ggplot2` you create a mapping between *x*-axis and *y*-axis variables. So, to create a plot of the `quakes` data, you map `quakes$long` to the *x*-axis and `quakes$lat` to the *y*-axis:

```
> ggplot(quakes, aes(x = long, y = lat)) + geom_point()
```

This creates Figure 18-3.

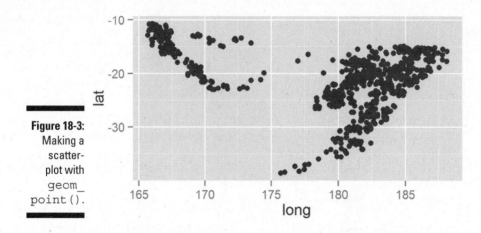

Figure 18-3:
Making a
scatter-
plot with
`geom_point()`.

Creating line charts

To create a line chart, you use the `geom_line()` function. You use this function in a very similar way to `geom_point()`; the difference is that `geom_line()` draws a line between consecutive points in your data.

This type of chart is useful for time series data in data frames, such as the population data in the built-in dataset `longley` (see Chapter 17). To create a line chart of unemployment figures, you use the following:

```
> ggplot(longley, aes(x = Year, y = Unemployed)) + geom_line()
```

This creates Figure 18-4.

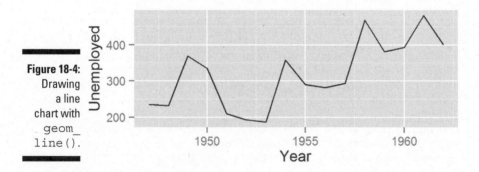

Figure 18-4:
Drawing
a line
chart with
`geom_line()`.

You can use either `geom_line()` or `geom_path()` to create a line drawing in `ggplot2`. The difference is that `geom_line()` first orders the observations according to x-value, whereas `geom_path()` draws the observations in the order found in the data.

Sussing Stats

After data, mapping, and geoms, the fourth element of a ggplot2 layer describes how the data should be summarized. In ggplot2, you refer to this statistical summary as a *stat*.

One very convenient feature of ggplot2 is its range of functions to summarize your data in the plot. This means that you often don't have to pre-summarize your data. For example, the height of bars in a histogram indicates how many observations of something you have in your data. The statistical summary for this is to count the observations. Statisticians refer to this process as *binning*, and the default stat for geom_bar() is stat_bin().

Analogous to the way that each geom has an associated default stat, each stat also has a default geom. Table 18-2 shows some useful stat functions, their effects, and their default geoms.

So, this begs the question: How do you decide whether to use a geom or a stat? In theory it doesn't matter whether you choose the geom or the stat first. In practice, however, it often is intuitive to start with a type of plot first — in other words, specify a geom. If you then want to add another layer of statistical summary, use a stat.

Figure 18-2, earlier in this chapter, is an example of this. In this plot, you used the same data to first create a scatterplot with geom_point(), and then you added a smooth line with stat_smooth().

Next, we take a look at some practical examples of using stat functions.

Table 18-2	Some Useful Stats and Default Geoms	
Stat	*Description*	*Default Geom*
stat_bin()	Counts the number of observations in bins.	geom_bar()
stat_smooth()	Creates a smooth line.	geom_line()
stat_sum()	Adds values.	geom_point()
stat_identity()	No summary. Plots data as is.	geom_point()
stat_boxplot()	Summarizes data for a box-and-whisker plot.	geom_boxplot()

Binning data

You've already seen how to use `stat_bin()` to summarize your data into bins, because this is the default stat of `geom_bar()`. This means that the following two lines of code produce identical plots:

```
> ggplot(quakes, aes(x = depth)) + geom_bar(binwidth = 50)
> ggplot(quakes, aes(x = depth)) + stat_bin(binwidth = 50)
```

Your plot should be identical to Figure 18-2.

Smoothing data

The `ggplot2` package also makes it very easy to create regression lines through your data. You use the `stat_smooth()` function to create this type of line.

The interesting thing about `stat_smooth()` is that it makes use of local regression by default. R has several functions that can do this, but `ggplot2` uses the `loess()` function for local regression. This means that if you want to create a linear regression model (as in Chapter 15), you have to tell `stat_smooth()` to use a different smoother function. You do this with the `method` argument.

To illustrate the use of a smoother, start by creating a scatterplot of unemployment in the `longley` dataset:

```
> p <- ggplot(longley, aes(x = Year, y = Employed)) + geom_point()
> p
```

Next, add a smoother. This is as simple as adding `stat_smooth()` to your line of code.

```
> p + stat_smooth()
```

Your graphic should look like the plot to the left of Figure 18-5.

Sometimes, `ggplot2` generates messages with extra tips and information. As long as you don't see warning or error, you can safely ignore these messages. In this case, `stat_smooth()` tells you that the default smoother is a method called *loess* (local smoothing). The message also says you can use alternative smoothing methods.

Finally, use `stat_smooth()` to fit and plot a linear regression model. You do this by adding the argument `method="lm"`:

```
> p + stat_smooth(method = "lm")
```

Your graphic should now look like the plot to the right in Figure 18-5.

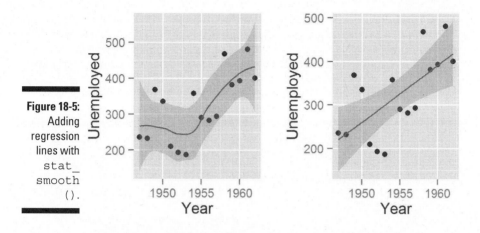

Figure 18-5:
Adding
regression
lines with
stat_
smooth
().

Doing nothing with identity

Sometimes you don't want ggplot2 to summarize your data in the plot.
This usually happens when your data is already pre-summarized or when
each line of your data frame has to be plotted separately. In these cases, you
want to tell ggplot2 to do nothing at all, and the stat to do this is stat_
identity(). You probably noticed in Table 18-1 that stat_identity is the
default statistic for points and lines.

Adding Facets, Scales, and Options

In addition to data, geoms, and stats, the full specification of a ggplot2
includes facets and scales. You've encountered facets in Chapter 17 — these
allow you to visualize different subsets of your data in a single plot. Scales
include not only the *x*-axis and *y*-axis, but also any additional keys that
explain your data (for example, when different subgroups have different
colors in your plot).

Adding facets

To illustrate the use of facets, you may want to replicate some of the faceted
plots of the dataset mtcars that you encountered in Chapter 17.

To make the basic scatterplot of fuel consumption against performance, use
the following:

```
> p <- ggplot(mtcars, aes(x = hp, y = mpg)) + geom_point()
> p
```

Then, to add facets, use the function `facet_grid()`. This function allows you to create a two-dimensional grid that defines the facet variables. You write the argument to `facet_grid()` as a formula of the form `rows ~ columns`. In other words, a tilde (~) separates the row variable from the column variable.

To illustrate, add facets with the number of cylinders as the columns. This means your formula is `~cyl`. Notice that because there are no rows as facets, there is nothing before the tilde character:

```
> p + stat_smooth(method = "lm") + facet_grid(~ cyl)
```

Your graphic should look like Figure 18-6.

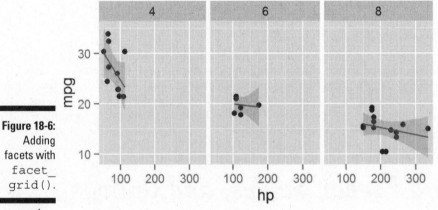

Figure 18-6: Adding facets with `facet_grid()`.

Similar to `facet_grid()`, you also can use the `facet_wrap()` function to wrap one dimension of facets to fill the plot grid.

Changing options

In `ggplot2`, you also can take full control of your titles, labels, and all other plot parameters.

To add *x*-axis and *y*-axis labels, you use the functions `xlab()` and `ylab()`.

Working with scales

In ggplot2, scales control the way your data gets mapped to your geom. In this way, your data is mapped to something you can see (for example, lines, points, colors, position, or shapes).

The ggplot2 package is extremely good at selecting sensible default values for your scales. In most cases, you don't have to do much to customize your scales. However, ggplot2 has a wide range of very sophisticated functions and settings to give you fine-grained control over your scale behavior and appearance.

In the following example, you map the column mtcars$cyl to both the shape and color of the points. This creates two separate, but overlapping, scales: One scale controls shape, while the second scale controls the color of the points:

```
> p <- ggplot(mtcars, aes(x = hp, y = mpg)) +
+    geom_point(aes(shape = factor(cyl), colour = factor(cyl)))
```

The name of a scale defaults to the name of the variable that gets mapped to it. In this case, you map factor(cyl) to the scale. To change the appearance of a scale, you need to add a scale function to your plot. The specific scale function you use is dependent on the type of scale, but in this case, you have a shape scale with discrete values, so you use the scale_shape_discrete() function. You also have a color scale with discrete value, so you can control that with scale_colour_discrete(). To change the name that appears in the legend of the plot, you need to specify the argument name to these scales. For example, change the name of the legend to "Cylinders" by setting the argument name = "Cylinders":

```
> p +
+    scale_shape_discrete(name = "Cylinders") +
+    scale_colour_discrete(name = "Cylinders")
```

Similarly, to change the x-axis scale, you would use scale_x_continuous().

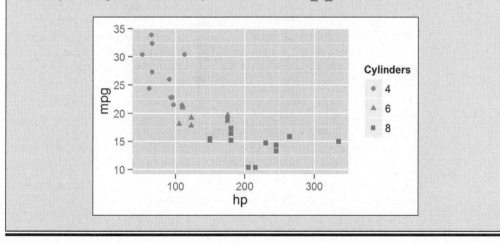

To add a main title, you use the function `ggtitle()`:

```
> ggplot(mtcars, aes(x = hp, y = mpg)) + geom_point(color = "red") +
+     xlab("Performance (horse power)") +
+     ylab("Fuel consumption (mpg)") +
+     ggtitle("Motor car comparison")
```

Your graphic should look like Figure 18-7.

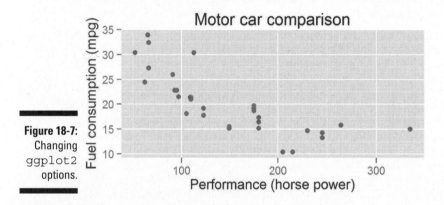

Figure 18-7:
Changing
`ggplot2`
options.

Getting More Information

In this chapter, we give you only a glimpse of the incredible power and variety at your fingertips when you use `ggplot2`. No doubt you're itching to do more. Here are a few resources that you may find helpful:

- ✔ The `ggplot2` website (`http://ggplot2.org/`) contains help and code examples for all the geoms, stats, facets, and scales. This site is an excellent resource. Because it also contains images of each plot, it's even easier to use than the built-in R Help.

- ✔ Hadley Wickham, the author of `ggplot2`, also wrote an excellent book that describes how to use `ggplot2` in a very easy and helpful way. The book is called *ggplot2: Elegant Graphics for Data Analysis,* and you can find its website at `http://ggplot2.org/book/`. At this website, you also can find some sample chapters that you can read free of charge.

- ✔ At `https://github.com/hadley/ggplot2/wiki`, you can find the `ggplot2` wiki, which is actively maintained and contains links to all kinds of useful information.

Part VI
The Part of Tens

In this part . . .

✔ Ten ways to replace Microsoft Excel.

✔ Ten things you really need to know about working with packages.

✔ Visit `www.dummies.com` for great Dummies content online.

Chapter 19

Ten Things You Can Do in R That You Would've Done in Microsoft Excel

In This Chapter

▶ Performing calculations and other operations on data

▶ Going beyond what you can do in a spreadsheet

The spreadsheet is probably one of the most widely used PC applications — and for good reason: Spreadsheets make it very easy to perform calculations and other operations on tabular data. But spreadsheets pose some risks as well: They're easy to corrupt and very difficult to debug.

The good news is, you can use R to do many of the same things you used to do in spreadsheets. In R, you use data frames to represent tabular data. R has many functions, operators, and methods that allow you to manipulate and calculate on data frames. This means that you can do just about anything (and more) in R that you would've done in Microsoft Excel, LibreOffice Calc, or your favorite spreadsheet application.

In this chapter, we offer some pointers on functions that you can explore in R, most of which are covered earlier in this book. In most cases, we provide some sample code but not the results. Try these examples yourself, and remember to use the R Help documentation to find out more about these functions.

Adding Row and Column Totals

One task that you may frequently do in a spreadsheet is calculating row or column totals. The easiest way to do this is to use the functions `rowSums()` and `colSums()`. Similarly, use `rowMeans()` and `colMeans()` to calculate means.

Try it on the built-in dataset `iris`. First, remove the fifth column, because it contains text that describes the species of iris:

```
> iris.num <- iris[, -5]
```

Then calculate the sum and mean for each column:

```
> colSums(iris.num)
> colMeans(iris.num)
```

These two functions are very convenient, but you may want to calculate some other statistic for each column or row. There's an easy way of traversing rows or columns of an array or data frame: the `apply()` function (see Chapter 13). For example, getting the minimum of a column is the same as applying a `min()` function to the second dimension of your data:

```
> apply(iris.num, 2, min)
> apply(iris.num, 2, max)
```

 The `apply()` function is ideal when your data is in an array and will apply happily over both rows and columns. For the special case where your data is in a data frame and you want to get column summaries, you're better off using `sapply()` rather than `apply()`. So, to get your `iris` column summaries, try this instead:

```
> sapply(iris.num, min)
> sapply(iris.num, max)
```

Formatting Numbers

When you produce reports, you want your numbers to appear all nicely formatted. For example, you may want to align numbers on the decimal points or specify the width of the column. Or you may want to print your number with a currency symbol ($100.00) or append a percentage sign to it (35.7%).

You can use `format()` to turn your numbers into pretty text, ready for printing. This function takes a number of arguments to control the format of your result. Here are a few:

- ✔ **trim:** A logical value. If FALSE, it adds spaces to right-justify the result. If TRUE, it suppresses the leading spaces.
- ✔ **digits:** How many significant digits of numeric values to show.
- ✔ **nsmall:** The minimum number of digits after the decimal point.

In addition, you control the format of the decimal point with decimal.mark, the mark between intervals *before* the decimal point with big.mark, and the mark between intervals *after* the decimal point with small.mark.

For example, you can print the number 12345.6789 with a comma as decimal point, spaces as the big mark, and dots as the small mark:

```
> format(12345.6789, digits = 9, decimal.mark = ",",
+        big.mark = " ", small.mark = ".", small.interval = 3)
[1] "12 345,678.9"
```

As a more practical example, to calculate the means of some columns in mtcars and then print the results with two digits after the decimal point, use the following:

```
> x <- colMeans(mtcars[, 1:4])
> format(x, digits = 2, nsmall = 2)
      mpg        cyl       disp         hp
" 20.09" "  6.19" "230.72" "146.69"
```

Notice that the result is no longer a number but a text string. So, be careful when you use number formatting — this should be the last step in your reporting workflow.

If you're familiar with programming in languages similar to C or C++, then you also may find the sprintf() function useful, because sprintf() is a wrapper around the C printf() function. This wrapper allows you to paste your formatted number directly into a string.

Here's an example of converting numbers into percentages:

```
> x <- seq(0.5, 0.55, 0.01)
> sprintf("%.1f %%", 100 * x)
[1] "50.0 %" "51.0 %" "52.0 %" "53.0 %" "54.0 %" "55.0 %"
```

This bit of magic should be familiar to C programmers, but for the rest of us, this is what it does: The first argument to sprintf() indicates the format — in this case, "%.1f %%". The format argument uses special literals that indicate that the function should replace this literal with a variable and apply some formatting. The literals always start with the % symbol. So, in this case, %.1f means to format the first supplied value as a fixed point value with one digit after the decimal point, and %% is a literal that means print a %.

To format some numbers as currency — in this case, U.S. dollars — use:

```
> set.seed(1)
> x <- 1000 * runif(5)
> sprintf("$ %3.2f", x)
[1] "$ 265.51" "$ 372.12" "$ 572.85" "$ 908.21" "$ 201.68"
```

As you saw earlier, the literal %3.2f means to format the value as a fixed point value with three digits before the decimal and two digits after the decimal.

The sprintf() function is a lot more powerful than that: It gives you an alternative way of pasting the value of any variable into a string:

```
> stuff <- c("bread", "cookies")
> price <- c(2.1, 4)
> sprintf("%s cost $ %3.2f ", stuff, price)
[1] "bread cost $ 2.10 "   "cookies cost $ 4.00 "
```

What happens here is that, because you supplied two vectors (each with two elements) to sprintf(), your result is a vector with two elements. R cycles through the elements and places them into the sprintf() literals. Thus, %s (indicating format the value as a string) gets the value "bread" the first time and "cookies" the second time.

You can do everything with paste() and format() that you can do with sprintf(), so you don't really ever need to use it. But when you do, it can simplify your code.

Sorting Data

To sort data in R, you use the sort() or order() functions (see Chapter 13).

To sort the data frame mtcars in increasing or decreasing order of the column hp, use:

```
> with(mtcars, mtcars[order(hp), ])
> with(mtcars, mtcars[order(hp, decreasing = TRUE), ])
```

Making Choices with If

Spreadsheets give you the ability to perform all kinds of "What if?" analyses. One way of doing this is to use the if() function in a spreadsheet.

R also has the `if()` function, but it's mostly used for flow control in your scripts. Because you typically want to perform a calculation on an entire vector in R, it's usually more appropriate to use the `ifelse()` function (see Chapter 9).

Here's an example of using `ifelse()` to identify cars with high fuel efficiency in the dataset `mtcars`:

```
> mtcars <- transform(mtcars,
+                     mpgClass = ifelse(mpg < mean(mpg), "Low", "High"))
> mtcars[mtcars$mpgClass == "High", ]
```

Calculating Conditional Totals

Something else that you probably did a lot in Excel is calculating conditional sums and counts with the functions `sumif()` and `countif()`.

You can do the same thing in one of two ways in R:

- ✔ Use `ifelse()` (see the preceding section).
- ✔ Simply calculate the measure of interest on a subset of your data.

Say you want to calculate a conditional mean of fuel efficiency in `mtcars`. You do this with the `mean()` function. Now, to get the fuel efficiency for cars either side of a threshold of 150 horsepower, try the following:

```
> with(mtcars, mean(mpg))
[1] 20.09062
> with(mtcars, mean(mpg[hp < 150]))
[1] 24.22353
> with(mtcars, mean(mpg[hp >= 150]))
[1] 15.40667
```

Counting the number of elements in a vector is the same as asking about its length. This means that the Excel function `countif()` has an R equivalent in `length()`:

```
> with(mtcars, length(mpg[hp > 150]))
[1] 13
```

Transposing Columns or Rows

Sometimes you need to transpose your data from rows to columns or vice versa. In R, the function to transpose a matrix is t():

```
> x <- matrix(1:12, ncol = 3)
> x
     [,1] [,2] [,3]
[1,]    1    5    9
[2,]    2    6   10
[3,]    3    7   11
[4,]    4    8   12
```

To get the transpose of a matrix, use t():

```
> t(x)
     [,1] [,2] [,3] [,4]
[1,]    1    2    3    4
[2,]    5    6    7    8
[3,]    9   10   11   12
```

You also can use t() to transpose data frames, but be careful when you do this. The result of a transposition is always a matrix (or array). Because arrays always have only one type of variable, such as numeric or character, the variable types of your results may not be what you expect.

Notice that the transposition of mtcars is a character array:

```
> t(mtcars[1:4, ])
         Mazda RX4 Mazda RX4 Wag Datsun 710 Hornet 4 Drive
mpg      "21.0"    "21.0"        "22.8"     "21.4"
cyl      "6"       "6"           "4"        "6"
disp     "160"     "160"         "108"      "258"
hp       "110"     "110"         " 93"      "110"
drat     "3.90"    "3.90"        "3.85"     "3.08"
wt       "2.620"   "2.875"       "2.320"    "3.215"
qsec     "16.46"   "17.02"       "18.61"    "19.44"
vs       "0"       "0"           "1"        "1"
am       "1"       "1"           "1"        "0"
gear     "4"       "4"           "4"        "3"
carb     "4"       "4"           "1"        "1"
mpgClass "High"    "High"        "High"     "High"
```

Finding Unique or Duplicated Values

To identify all the unique values in your data, use the unique() function. Try finding the unique values of number of cylinders in mtcars:

```
> unique(mtcars$cyl)
[1] 6 4 8
```

Sometimes you want to know which values of your data are duplicates. Depending on your situation, those duplicates will be valid, but sometimes duplicate entries may indicate data-entry problems.

The function to identify duplicate entries is duplicated(). In the built-in dataset iris, there is a duplicated row in line 143. Try it yourself:

```
> dupes <- duplicated(iris)
> head(dupes)
[1] FALSE FALSE FALSE FALSE FALSE FALSE
> which(dupes)
[1] 143
> iris[dupes, ]
    Sepal.Length Sepal.Width Petal.Length Petal.Width  Species
143          5.8         2.7          5.1         1.9 virginica
```

Because the result of duplicated() is a logical vector, you can use it as an index to remove rows from your data. To do this, use the negation operator — the exclamation point (as in !dupes):

```
> iris[!dupes, ]
> nrow(iris[!dupes, ])
[1] 149
```

Working with Lookup Tables

In a spreadsheet application like Excel, you can create lookup tables with the functions vlookup or a combination of index and match.

In R, it may be convenient to use merge() (see Chapter 13) or match(). The match() function returns a vector with the positions of elements that match your lookup value.

For example, to find the location of the element `"Toyota Corolla"` in the row names of `mtcars`, try the following:

```
> index <- match("Toyota Corolla", rownames(mtcars))
> index
[1] 20
> mtcars[index, 1:4]
                mpg cyl disp hp
Toyota Corolla 33.9   4 71.1 65
```

You can see that the index position is 20, and that the 20th row is indeed the row you're looking for.

Working with Pivot Tables

In Excel, pivot tables are a useful tool for manipulating and analyzing data.

For simple tables in R, you can use the `tapply()` function to achieve similar results. Here's an example of using `tapply()` to calculate mean `hp` for cars with different numbers of cylinders and gears:

```
> with(mtcars, tapply(hp, list(cyl, gear), mean))
          3      4     5
4  97.0000   76.0 102.0
6 107.5000  116.5 175.0
8 194.1667     NA 299.5
```

For slightly more complex tables — that is, tables with more than two cross-classifying factors — use the `aggregate()` function:

```
> aggregate(hp ~ cyl + gear + am, mtcars, mean)
   cyl gear am        hp
1    4    3  0  97.00000
2    6    3  0 107.50000
3    8    3  0 194.16667
4    4    4  0  78.50000
5    6    4  0 123.00000
6    4    4  1  75.16667
7    6    4  1 110.00000
8    4    5  1 102.00000
9    6    5  1 175.00000
10   8    5  1 299.50000
```

If you frequently work with tables in Excel, you should definitely explore the packages `dplyr` and `tidyr` that are available on CRAN at `http://cran.r-project.org/web/packages/dplyr` and `http://cran.r-project.org/web/packages/tidyr`, respectively. These packages provide a number of functions for common data manipulation problems.

Using the Goal Seek and Solver

One very powerful feature of Excel is that it has a very easy-to-use solver that allows you to find minimum or maximum values for functions given some constraints.

A very large body of mathematics aims to solve optimization problems of all kinds. In R, the `optimize()` function provides one fairly simple mechanism for optimizing functions.

Imagine you're the sales director of a company and you need to set the best price for your product. In other words, find the price of a product that maximizes revenue.

In economics, a simple model of pricing states that people buy less of a given product when the price increases. Here's a very simple function that has this behavior:

```
> sales <- function(price) { 100 - 0.5 * price }
```

Expected revenue is then simply the product of price and expected sales:

```
> revenue <- function(price) { price * sales(price) }
```

You can use the `curve()` function to plot continuous functions. This takes a function as input and produces a plot. Try to plot the behavior of sales and revenue using the `curve()` function, varying price from $50 to $150:

```
> oldpar <- par(mfrow = c(1, 2), bty = "l")
> curve(sales, from = 50, to = 150, xname = "price", main = "Sales")
> curve(revenue, from = 50, to = 150, xname = "price", main = "Revenue")
> par(oldpar)
```

Your results should look similar to Figure 19-1.

You have a working model of sales and revenue. From the figure, you can see immediately that there is a point of maximum revenue. Next, use the R function `optimize()` to find the value of that maximum. To use `optimize()`, you need to tell it which function to use (in this case, `revenue()`),

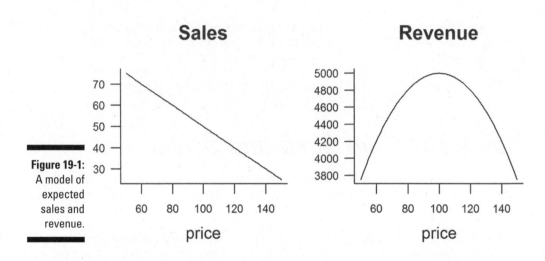

Figure 19-1:
A model of
expected
sales and
revenue.

as well as the interval (in this case, prices between 50 and 150). By default,
`optimize()` searches for a minimum value, so in this case you have to tell
it to search for maximum value:

```
> optimize(revenue, interval = c(50, 150), maximum = TRUE)
$maximum
[1] 100

$objective
[1] 5000
```

And there you go. Charge a price of $100, and expect to get $5,000 in revenue.

The Excel Solver uses the Generalized Reduced Gradient Algorithm for
optimizing nonlinear problems (http://support.microsoft.com/
kb/214115). The R function `optimize()` uses a combination of golden
section search and successive parabolic interpolation, which clearly is not
the same thing as the Excel Solver. Fortunately, a large number of packages
provide various different algorithms for solving optimization problems. In
fact, there is a special task view on CRAN for optimization and mathematical programming. Go to http://cran.r-project.org/web/views/
Optimization.html to find out more than you ever thought you wanted
to know!

Chapter 20

Ten Tips on Working with Packages

*O*ne of the very attractive features of R is that it contains a large collection of third-party *packages* (collections of functions in a well-defined format). To get the most out of R, you need to understand where to find additional packages, how to download and install them, and how to use them.

In this chapter, we consolidate some of the things we cover earlier in the book and give you ten tips on working with packages.

 Many other software languages have concepts that are similar to R packages. Sometimes these are referred to as "libraries." However, in R, a library is the folder on your hard disk (or USB stick, network, DVD, or whatever you use for permanent storage) where your packages are stored.

Poking Around the Nooks and Crannies of CRAN

The Comprehensive R Archive Network (CRAN; http://cran.r-project.org) is a network of web servers around the world where you can find the R source code, R manuals and documentation, and contributed packages.

CRAN isn't a single website; it's a collection of web servers, each with an identical copy of all the information on CRAN. Thus, each web server is called a *mirror*. The idea is that you choose the mirror that is located nearest to where you are, which reduces international or long-distance Internet traffic. You can find a list of CRAN mirrors at http://cran.r-project.org/mirrors.html.

RGui and RStudio allow you to set the location of your nearest CRAN mirror directly in the application. For example, in the Windows RGui, you can find this option by choosing Packages⇨Set CRAN mirror. In RStudio, you can find this option by choosing Tools⇨Global Options⇨Packages⇨CRAN mirror.

Regardless of which R interface you use, you can permanently save your preferred CRAN mirror (and other settings) in a special file called .RProfile, located in the user's home directory or the R startup directory. For example, to set the Imperial College, UK mirror as your default CRAN mirror, include this line in your .RProfile:

```
options("repos" = c(CRAN = "http://cran.ma.imperial.ac.uk/"))
```

For more information, see Appendix A.

Finding Interesting Packages

At the beginning of 2015, there were more than 6,000 packages on CRAN. That means finding a package for your task at hand may seem difficult.

Fortunately, a handful of volunteer experts have collated some of the most widely used packages into curated lists. These lists are called *CRAN task views,* and you can view them at http://cran.r-project.org/web/views/. You can find task views for empirical finance, statistical genetics, machine learning, statistical learning, and many other fascinating topics.

Each package has its own web page on CRAN. Say, for example, you want to find a package to do high-quality graphics. If you followed the link to the graphics task view, http://cran.r-project.org/web/views/Graphics.html, you may notice a link to the ggplot2 package, http://cran.r-project.org/web/packages/ggplot2/index.html. On the web page for a package, you find a summary, information about the packages that are used, a link to the package website (if such a site exists), and other useful information.

Installing Packages

To install a package use the `install.packages()` function. This simple command downloads the package from a specified repository (by default, CRAN) and installs it on your machine:

```
> install.packages("fortunes")
```

Note that the argument to `install.packages()` is a character string. In other words, remember the quotes around the package name!

In RGui, as well as in RStudio, you find a menu command to do the same thing:

- ✔ In RGui, choose Packages⇨Install package(s).
- ✔ In RStudio, choose Tools⇨Install packages. . . .

Loading Packages

To load a package, you use the `library()` or `require()` function. These functions are identical in their effects, but they differ in the return value:

- ✔ **`library()`:** Invisibly returns a list of packages that are attached, or stops with an error if the package is not on your machine.
- ✔ **`require()`:** Returns TRUE if the package was successfully attached and FALSE if not.

The R documentation suggests that `library()` is the preferred way of loading packages in scripts, while `require()` is preferred inside functions and packages.

So, after installing the package `fortunes` you load it like this:

```
> library("fortunes")
```

Note that you don't have to quote the name of the package in the argument of `library()`, but it is good practice to always quote the package name.

Although it is possible to unload a package within an R session by using the `detach()` function, in practice it usually is much easier to simply restart your R session.

Reading the Package Manual and Vignette

After installing and loading a new package, a good starting point is to read the package manual. The package manual is a collection of all functions and other package documentation. You can access the manual in two ways. The first way is to use the help argument to the library() function:

```
> library(help = "fortunes")
```

The second way is to find the manual on the package website. If you point your browser window to the CRAN page for the fortunes package (http://cran.r-project.org/web/packages/fortunes/), you'll notice a link to the manual toward the bottom of the page (http://cran.r-project.org/web/packages/fortunes/fortunes.pdf).

Whichever approach you choose, the result is a PDF document containing the package manual.

Some package authors also write one or more *vignettes,* documents that illustrate how to use the package. A vignette typically shows some examples of how to use the functions and how to get started. The key thing is that a vignette illustrates how to use the package with R code and output, just like this book.

To read the vignette for the fortunes package, try the following:

```
> vignette("fortunes")
```

Updating Packages

Most package authors release improvements to their packages from time to time. To ensure that you have the latest version, use update.packages():

```
> update.packages()
```

This function connects to CRAN (by default) and checks whether there are updates for all the packages that you've installed on your machine. If there are, it asks you whether you want to update each package, and then downloads the code and installs the new version.

If you add `update.packages(ask = FALSE)`, R updates all out-of-date packages in the current library location, without prompting you. Also, you can tell `update.packages()` to look at a repository other than CRAN by changing the `repos` argument. If the `repos` argument points to a file on your machine (or network), R installs the package from this file.

Both RGui and RStudio have menu options that allow you to update the packages:

- ✓ In RGui, choose Packages⇨Update package(s).
- ✓ In RStudio, choose Tools⇨Check for Package Updates. . . .

Both applications allow you to graphically select packages to update.

Forging Ahead with R-Forge

Although not universally true, packages on CRAN tend to have some minimum level of maturity. For example, to be accepted by CRAN, a package needs to pass a basic minimum set of requirements.

So, where do packages live that are in the development cycle? Quite often, they live at R-Forge (`http://r-forge.r-project.org/`). R-Forge gives developers a platform to develop and test their R packages. For example, R-Forge offers

- ✓ A build and check system on Windows and Linux operating systems (Mac OSX is not supported)
- ✓ Version control
- ✓ Bug-report systems
- ✓ Backup and administration

To install a project from R-Forge, you also use the `install.packages()` function, but you have to specify the `repos` argument. For example, to install the development version of the package `data.table`, try the following:

```
> install.packages("data.table", repos = "http://R-Forge.R-project.org")
```

Although R-Forge doesn't have a build and check system for Mac OSX specifically, Mac users can install and use packages from R-Forge by installing the source package. A source package is a compressed file (with a `.tar.gz` file extension) containing all source code for the package. On installing, Linux builds the package "from source". As OSX is built on Linux, in most cases you can install the source package on Mac OSX without any additional tools. You find more information in the FAQ for Mac at this website:

```
http://cran.r-project.org/bin/macosx/RMacOSX-FAQ.html
```

Getting packages from github

In recent years, many developers have started to use github (https://github.com) as a code development site. Although github does not offer any of the R-specific features of CRAN or R-Forge, sometimes code is easier to share by using github. So you may occasionally get instructions to install a package directly from github.

On the Linux and Mac OSX operating systems, installing packages from github is comparatively easy. However, on Windows you also must first install RTools (a set of compilers and other tools to build packages from source). To install RTools on a Windows machine, carefully follow the instructions at http://cran.r-project.org/bin/windows/Rtools.

For example, to install the rfordummies package, associated with this book, directly from github, try:

```
> install.packages("devtools")
> library("devtools")
> install_github("andrie/rfordummies")
```

The rfordummies package also is available on CRAN, so it is much simpler to install using

```
> install.packages("rfordummies", dependencies = TRUE)
```

You can find out more about the rfordummies package in Appendix B.

Conducting Installations from BioConductor

BioConductor is a repository of R packages and software, a collection of tools that specializes in analysis of genomic and related data.

BioConductor has its own sets of rules for developers. For example, to install a package from BioConductor you have to source a script from its server:

```
> source("http://bioconductor.org/biocLite.R")
```

Then you can use the `biocLite()` function to install packages from BioConductor. If you don't supply an argument, you just install the necessary base packages from the BioConductor project. You can find all the information you need at `www.bioconductor.org`.

BioConductor extensively uses object-orientation programming with S4 classes. Object orientation and its implementation as S4 classes is an advanced R topic — one we don't discuss in this book. If you're interested, you find more information in Hadley Wickham's great online book *Advanced R* at `http://adv-r.had.co.nz/OO-essentials.html`.

Reading the R Manual

The "R Installation and Administration" manual (`http://cran.r-project.org/doc/manuals/R-admin.html`) is a comprehensive guide to the installation and administration of R. Chapter 6 of this manual contains all the information you need about working with packages. You can find it at `http://cran.r-project.org/doc/manuals/R-admin.html#Add_002don-packages`.

Appendix A

Installing R and RStudio

● ●

*B*efore you use R, of course, you first have to install R. Although you can use the built-in code editor, you may also want to install an editor with more functionality. Because RStudio runs on all platforms and is integrated nicely with R, we also discuss the installation of RStudio on your system.

In this appendix, we don't have enough space to provide installation instructions for every possible operating system. You can find that information on the R and RStudio websites:

- ✔ **R:** www.r-project.org
- ✔ **RStudio:** www.rstudio.com

If you use Linux, depending on the distribution you use, you may find that R comes with the operating system and doesn't require a separate installation.

Installing and Configuring R

Installing R isn't difficult, but tweaking it to fit your own needs requires a bit of explanation.

We can't cover all possibilities here, so be sure to read the information on the R website for more insight on how to install and configure the software.

Installing R

You can find the installation files and all information about installation on one of the mirror sites of the Comprehensive R Archive Network (CRAN; for example, http://cran.rstudio.com/). Select the download link for your operating system, which will take you to the download site for the latest version of R.

You can find detailed installation instructions in the R Installation and Administration manual on CRAN (`http://cran.r-project.org/doc/manuals/R-admin.html`). For Windows, you take the following steps:

1. **Go to CRAN, click Download R for Windows, click Base, and download the installer for the latest R version.**

2. **Right-click the installer file and select Run as Administrator from the pop-up menu.**

3. **Select the language to be used during installation.**

   This doesn't change the language used by R; all messages and Help files are determined by the language settings of your computer.

4. **Follow the instructions of the installer.**

   You can safely use the default settings and just keep clicking Next until R starts installing.

R exists in a 32-bit and 64-bit version. If you have a 64-bit Windows version, you can easily install both versions next to each other. (The installer will do this by default.) For other systems, you can find more information in the R Installation and Administration manual. The 32-bit version of R is perfectly fine — sometimes it's even a bit faster than the 64-bit version. You need the 64-bit version only if you require more work memory than the 32-bit version can handle. (On Windows, the maximum is about 3GB for a 32-bit system.)

If you want to be able to personalize your R installation as explained here, you should install R outside the `Program Files` folder (for example, in `C:\R\`). This way, you avoid trouble with the default Windows folder protection.

Mac OS X and Linux users especially need to read the instructions on the CRAN site carefully. R can be installed on all systems, but depending on your version of OS X or Linux, you may need to follow special procedures to install R successfully. Not following these procedures could harm your system.

Configuring R

Apart from accepting the options in the installation procedure, you can change a number of startup options by adapting the `Rprofile.site` file. This file is located inside the installation directory, in the subfolder `.../R-n.n.n/etc` (for example, `.../R-3.2.0/etc`). The file is sourced by R at startup, so all R code in this file is carried out. The default installation of R contains a perfectly valid `Rprofile.site` file, so you have to change this only if you want to personalize your startup.

`Rprofile.site` is a normal text file, so you can edit it as you would any other text file. The file already contains some options that are commented out, so you get a good idea of what's possible when you open the file in a text editor (for example, Notepad). Be sure to check the Help page `?options` to get more information on all possible options.

You can personalize R further by adding a code file called `.Rprofile` to your personal home folder. You can find this folder from within R by setting the working directory to `"~/"`, like this:

```
> setwd("~/")
> getwd()
[1] "C:/Users/Joris FA Meys/Documents"
```

Adding an `.Rprofile` file isn't necessary, but R will always look for one, either in the folder from which you call R or in the user's home directory. Whereas an `Rprofile.site` file is linked to a specific installation of R, the `.Rprofile` file can differ for every user on the same system. If you update R, you can leave the `.Rprofile` file where it is and the new R version will automatically find it and apply the options you specified there. So, after updating R to the latest version, you have to adapt the `Rprofile.site` again only if you want to personalize it.

Functions you define or objects you create with code in `Rprofile.site` won't be visible if you use `ls()`, although you can use them without trouble. This also means you can't delete them easily from the workspace.

An `Rprofile.site` or `.Rprofile` file may look like the following example:

```
# Sample profile file
# Set CRAN mirror to a default location
options(repos = "http://cran.uk.r-project.org")
# R interactive prompt
options(prompt = "R: ")
# sets work directory back to original
go.home <- function() setwd("D:/MyWorkspace")
```

Using this file, R starts up with a different prompt (`R:` instead of `>`) and sets the mirror from the UK as the default mirror from which to install packages. You also define the `go.home()` function, which you can use at any point to set your working directory back to your home directory (`D:/MyWorkspace`, in this example).

Installing and Configuring RStudio

RStudio is a relatively new and shiny editor for R. It's our first choice for this book because it's easy to use, it has a good documentation website, it has very good support, and it incorporates R in a practical way. Of course, you're free to work with any code editor you like; in Chapter 2, we discuss some alternatives.

Installing RStudio

Installing RStudio is easy. Just follow these steps:

1. Go to `http://www.rstudio.com/products/RStudio/#Desk`.

2. **Click the Download RStudio Desktop button.**

3. **Select the installation file for your system.**

4. **Run the installation file.**

RStudio will be installed on your system. It normally detects your latest installed R version automatically. If you didn't do anything funky, you should be able to use R from within RStudio without extra configuration.

Configuring RStudio

You may want to use a different R version from the one RStudio detected. For example, you may want to use R in a 64-bit context. Or RStudio may not have recognized your installation of R. In that case, you can set which R version to use by choosing Tools⇨Global Options. . . to open the Global Options pane (see Figure A-1).

To change the R version, click the Change button. Then you can switch between the default 32-bit R installation and the 64-bit R installation (if installed), or you can choose a specific version of R. (RStudio lists all the versions it finds.)

If you click Browse, you can select the root directory for any R version you want to use. This folder normally looks something like `.../R/R-n.n.n`. If you select an R version that has both 32-bit and 64-bit builds, RStudio will ask you which build you want to use.

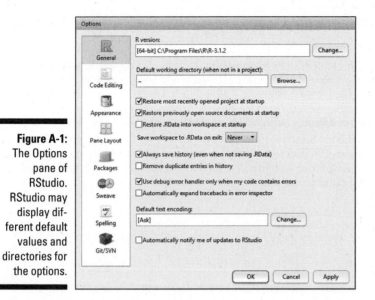

In the Options pane (refer to Figure A-1), you can tweak the behavior of R in RStudio. If you click the General icon in the left column, you can set a number of default options for R in RStudio:

- ✔ **Initial working directory:** You can set the default working directory R uses at startup.

- ✔ **Save workspace to .RData on exit:** Your options are Ask, Never, or Always. By default, RStudio asks you whether you want to save the workspace when you quit.

- ✔ **Restore .RData into workspace at startup:** Select this check box to let RStudio restore the workspace automatically at startup. RStudio will look for a saved workspace in the root folder or the default working directory.

- ✔ **Always save history (even when not saving .RData):** Select this check box to have RStudio always save the history. If you don't select this check box, RStudio doesn't save the history when you exit the program.

- ✔ **Remove duplicate entries in history:** If you select this check box, RStudio removes duplicate entries from your history when the history is saved.

The remaining icons in the options panel give you more possibilities for adapting the general appearance of RStudio, the appearance of the code editing tool or script window, and the general layout of the panes. For example, you can set the default CRAN mirror for installation of packages by clicking on the **Packages** icon in the Options panel. By default RStudio uses its own repository.

These settings work for R only from within RStudio. If you use R with another code editor or by itself, the settings in RStudio will have no effect.

As you can see in Figure A-1, the Options pane also offers you some more advanced options. RStudio evolved to a complete Integrated Development Environment (IDE) for R, and comes packed with tools for development of your own packages. We don't cover this advanced level of R programming in this book, but the RStudio website contains information on the possibilities for taking your R code to the next level.

Appendix B

The rfordummies Package

● ●

*Y*ou can find all of the example code of this book in an R package called rfordummies. You can install this package directly from CRAN and then use the package to view the code and run the examples in the book.

Using rfordummies

You install the rfordummies package directly from CRAN using the function install.packages(). If you add the argument dependencies = TRUE, you automatically install all the packages mentioned in the book, like this:

```
> install.packages("rfordummies", dependencies = TRUE)
```

You load the package by using the function library():

```
> library("rfordummies")
```

You can view the table of contents of the book by using the function toc():

```
> toc()
```

To view the chapter code for a specific chapter, use the function chxx(), where xx means the chapter number. For example, to view the code for Chapters 2 and 13, enter these functions:

```
> ch2()
> ch13()
```

To run the examples from a specific chapter, use the example() function. This function runs all the example code given on a certain help page. So to run all examples from Chapter 3, try:

```
> example("ch3", package = "rfordummies")
```

You can also find the data for the periodic table of elements (see Chapter 12) in a data frame called `elements`:

```
> ?elements
> head(elements)
```

For more information, check the help page `?rfordummies`.

Index

• *G* •

Authors' Acknowledgments

This book is possible only because of the tremendous support we had from our editorial team at Wiley. In particular, thanks to our project editors, Elizabeth Kuball (first edition) and Katie Mohr (second edition).

Thank you to our technical editor, Gavin Simpson, for his thorough reading and many helpful comments.

We wish to thank Patrick Burns, author of the R Inferno, for his enthusiastic support and very detailed review of the first edition.

Thank you to the R core team for creating R, for maintaining CRAN, and for your dedication to the R community in the form of mailing lists, documentation, and seminars. And thank you to the R community for creating thousands of useful packages, writing blogs, and answering questions.

In this book, we use several packages by Hadley Wickham, whose contribution of ggplot graphics and other helpful packages like dplyr continues to be an inspiration.

While writing this book we had very helpful support from a large number of contributors to the R tag at Stack Overflow. Thank you to James (JD) Long, David Winsemius, Ben Bolker, Joshua Ulrich, Barry Rowlingson, Roman Luštrik, Joran Elias, Dirk Eddelbuettel, Richie Cotton, Colin Gillespie, Simon Urbanek, Gabor Grotendieck, and everybody else who continue to make Stack Overflow a great resource for the R community.

From Andrie: It is not an exaggeration to say that this book was partly responsible for changing the course of my life. Learning R, contributing in the open source community and writing this book all contributed to employment at Revolution Analytics. I want to thank all my colleagues, in particular Derek McCrae Norton, David Smith and Joseph Rickert.

From Joris: Thank you to the professors and my colleagues at the Department of Mathematical Modeling, Statistics, and Bioinformatics at Ghent University (Belgium) for the insightful discussions and ongoing support I received while writing this book.

Publisher's Acknowledgments

Project Editor: Pat O'Brien

Technical Editor: Gavin Simpson

Editorial Assistant: Claire Brock

Sr. Editorial Assistant: Cherie Case

Project Coordinator: Kumar Chellappan

About the Authors

Andrie de Vries: Andrie started to use R in 2009 to analyze survey data, and he has been continually impressed by the ability of the open-source community to innovate and create phenomenal software. During 2009 he also started PentaLibra, a boutique market research and statistical analysis agency. After getting increasingly involved in the R community, he joined Revolution Analytics to help take R to enterprise customers, helping clients to deal with the challenges of data science and big data. To maintain equilibrium in his life, Andrie is studying and practicing yoga.

Joris Meys, MSc: Joris is a statistical consultant, R programmer and R lecturer at Ghent University (Belgium). After earning a master's degree in biology, he worked for six years in environmental research and management before starting an advanced master's degree in statistical data analysis. Joris writes packages for both specific projects and general implementation of methods developed in his department, and he is the maintainer of several packages on R-Forge. He has co-authored a number of scientific papers as a statistical expert. To balance science with culture, Joris spends most of his spare time playing saxophone in a couple of local bands.

Dedication

This book is for my wife, Annemarie, because of her encouragement, support, and patience. Also for my niece, Tanya, who is really good at math and kept reminding me of approaching deadlines! Finally, to my parents, for a lifetime of encouragement.

—Andrie de Vries

I dedicate this book to the most important women in my life. For my mother, because she made me the man I am. For Eva, because she loves the man I am. For Amelie, because her little smile melts my heart every time. And for Granny, because she rocks!

—Joris Meys